Lecture Notes in Mathematics

Edited by A. Dold and B. Eckmann

T0253765

931

David M. Arnold

Finite Rank Torsion Free Abelian Groups and Rings

Springer-Verlag
Berlin Heidelberg New York 1982

Author
David M. Arnold
Department of Mathematical Sciences
New Mexico State University
Las Cruces, NM 88003, USA

AMS Subject Classifications (1980): 20K15, 20K30, 20K40

ISBN 3-540-11557-9 Springer-Verlag Berlin Heidelberg New York
ISBN 0-387-11557-9 Springer-Verlag New York Heidelberg Berlin

Printing and binding: Beltz Offsetdruck, Hemsbach/Bergstr.
2141/3140-543210

INTRODUCTION

These notes contain a largely expository introduction to the theory of finite rank torsion free abelian groups developed since the publication of "Infinite Abelian Groups," Vol. II, L. Fuchs, in 1973. As reflected in Chapter XIII of that text, the subject consists of a satisfactory theory for direct sums of rank 1 groups due to R. Baer in 1937; a uniqueness of quasi-direct sum decompositions up to quasi-isomorphism due to B. Jónsson in 1959; a realization of subrings of finite dimensional Q-algebras as endomorphism rings due to A.L.S. Corner in 1963; a variety of pathological direct sum decompositions; and some apparently miscellaneous results largely relegated to the exercises.

Substantial progress has been made in the subject since 1973. Most notable are the stable range conditions proved by R.B. Warfield, near isomorphism as introduced by E.L. Lady, and the application of properties of subrings of finite dimensional Q-algebras to finite rank torsion free abelian groups via a Morita-like duality developed by E.L. Lady and the author. Consequently, some older results of R. Beaumont, R. Pierce, and J. Reid (c. 1960) involving subrings of finite dimensional Q-algebras gain new importance. Thus a systematic introduction to the theory of finite rank torsion free abelian groups and subrings of finite dimensional Q-algebras seems timely.

The theory of direct sums of rank-1 torsion free abelian groups has been combined with the theory of totally projective groups to characterize a class of mixed abelian groups (Warfield [7] and Hunter-Richman [1]). The category Walk, as discussed in Warfield [7], has been used to investigate mixed abelian groups. A secondary goal of these notes is to survey the known results for finite rank torsion free abelian groups with an eye towards eventual application to mixed groups of finite torsion free rank via the category Walk. Some progress along these lines is reported by Warfield [7]. Other potential applications include the study of mixed abelian groups of finite torsion free rank via the category Warf, as discussed in Arnold - Hunter - Richman [1], and valuated finite direct sums of torsion free cyclic groups, as discussed in a series of papers by E. Walker, F. Richman, R. Hunter, and the author. In particular, Rotman [1] shows that finite rank torsion free groups are characterized in terms of valuated finite direct sums of torsion free cyclic groups.

These notes were developed for a graduate course taught by the author as part of the Year of Algebra at the University of Connecticut during academic year 1978-1979. The students were assumed to have had a graduate course in algebra (fundamental concepts and classical theory of artinian rings are given in Section 0 and the exercises) but little or no exposure to finite rank torsion free abelian groups or subrings of finite dimensional Q-algebras.

Except for portions of Sections 0, 1, and 2 there is little overlap with the results proved in Fuchs [7], Vol. II. There are exercises at the end of each section, some of which are contributed by others as noted, devoted to an extension and elaboration of the results presented or of the requisite background material. No attempt has been made to state or prove results in maximum generality, but in most cases references are given for more general theorems.

Sections 1-4 include a classical introduction to the subject of finite rank torsion free abelian groups as well as some generalizations of type and applications (Richman [1] and Warfield [1]) in Section 1; properties of rank-2 groups in terms of their typeset (Beaumont-Pierce [2]) in Section 3; and characterizations of pure subgroups of finite rank completely decomposable groups (Butler [1]) in Section 4.

Generalizations of such topics as finite rank completely decomposable groups and Baer's Lemma are developed in Sections 5-6 as derived by Arnold-Lady [1] and Arnold-Hunter-Richman [1].

Section 7 includes a proof of the Krull-Schmidt Theorem in additive categories with Jónsson's quasi-decomposition theorem and some essential properties of near isomorphism, due to Lady [1], as corollaries.

Stable range conditions are considered in Section 8 (Warfield [5]) as well as cancellation and substitution properties (Warfield [5], Fuchs-Loonstra [2], and Arnold-Lady [1]), exchange properties (Warfield [5], Monk [1], Crawley-Jónsson [1]) and self-cancellation (Arnold [7]).

Sections 9-11 include an extensive introduction to the subject of subrings of finite dimensional Q-algebras, including a proof of the Jordan-Zassenhaus Theorem for Z-orders, derived in part from Reiner [1] and Swan-Evans [1]. The fact that the additive groups of such rings are finite rank torsion free is exploited to avoid completions in the derivation of the theory. Moreover, localization at primes of Z is consistently used instead of localization at prime ideals of more general domains.

The relationship between near isomorphism and genus class of lattices over orders is examined in Section 12. Classical properties of genus classes of lattices over orders are derived and used to develop properties of near isomorphism of finite rank torsion free abelian groups.

The structure of Grothendieck groups of finite rank torsion free abelian groups is considered in Section 13, as developed by Lady [2] and Rotman [2].

Section 14 includes characterizations of additive groups of subrings of finite dimensional Q-algebras, due to Beaumont-Pierce [1] and [3], including a proof of the Wedderburn principal theorem and a simplified proof of the analog for subrings of finite dimensional Q-algebras.

Several classes of groups are given in Section 15, providing an appropriate setting for the development of Murley groups (Murley [1]) and strongly homogeneous groups (Arnold [6]).

The author expresses his appreciation to the University of Connecticut Department of Mathematics for the invitation to participate in the Year of Algebra. Special thanks are due to R. Pierce, J. Reid, C. Vinsonhaler, W. Wickless,

R. Weigand, S. Weigand, T. Faticoni, and J. Chung for their encouragement, assistance, and extraordinary patience with the development of these notes. The author is also indebted to E.L. Lady for the opportunity to read an unpublished set of lecture notes on the subject, R. B. Warfield, Jr. for numerous discussions regarding the contents of these notes, and D. Boyer, A. Mader, M. Maher, K. Rangaswamy, T. Giovannitti, J. Moore, U. Albrecht and J. Stelzer for their helpful comments and suggestions. A special note of thanks is due to Lorna Schriver for her accurate and efficient typing of these notes as well as her remarkable ability to interpret the author's frequently illegible handwriting. Finally, any errors that remain are directly attributable to the author.

TABLE OF CONTENTS

§0. Notation and Preliminaries

It is assumed that rings have identities and that ring homomorphisms preserve identities. In particular, if R is a subring of S then $1_R = 1_S$.

Suppose that R is a ring, A is a right R-module and B is a left R-module. Then $A\otimes_R B$ is defined to be F/N where F is the free abelian group with elements of $A\times B$ as a basis and N is the subgroup of F generated by $\{(a_1 + a_2, b) - (a_1, b) - (a_2, b), (a, rb) - (ar, b), (a, b_1 + b_2) - (a, b_1) - (a, b_2) | a, a_1, a_2 \in A; b, b_1, b_2 \in A$ and $r \in R\}$. Write $a \otimes b$ for $(a, b) + N$ so that if $x \in A\otimes_R B$ then $x = \Sigma\, a_i\otimes b_i$ for some $a_i \in A$, $b_i \in B$. Then $\psi : A\times B \to A\otimes_R B$, defined by $\psi(a,b) = a \otimes b$ is an R bi-linear map. Furthermore, if G is an abelian group and $g : A\times B \to G$ is an R bi-linear map then there is a unique homomorphism $\psi : A\otimes_R B \to G$ with $\phi\psi = g$.

With the above properties, one can prove that

(i) $A\otimes_R (\oplus B_i) \simeq \oplus (A\otimes_R B_i)$;

(ii) If $f \in \mathrm{Hom}_R(A, A')$ and $g \in \mathrm{Hom}_R(B, B')$ then $f \otimes g : A\otimes_R B \to A'\otimes_R B'$, induced by $(f \otimes g)(a \otimes b) = f(a) \otimes g(b)$, is a homomorphism of abelian groups;

(iii) $A\otimes_R *$ and $*\otimes_R B$ are functors, e.g., if $f \in \mathrm{Hom}_R(B_1, B_2)$ and $g \in \mathrm{Hom}_R(B_2, B_3)$ then $(1_A\otimes g)(1_A\otimes f) = 1_A\otimes gf : A\otimes_R B_1 \to A\otimes_R B_3$ and $1\otimes 1 : A\otimes_R A \to A\otimes_R A$ is the identity homomorphism;

(iv) If $0 \to B_1 \to B_2 \to B_3 \to 0$ is an exact sequence of left R-modules then $A\otimes_R B_1 \to A\otimes_R B_2 \to A\otimes_R B_3 \to 0$ is an exact sequence of abelian groups and if $0 \to A_1 \to A_2 \to A_3 \to 0$ is an exact sequence of right R-module then $A_1\otimes_R B \to A_2\otimes_R B \to A_3\otimes_R B \to 0$ is an exact sequence of abelian groups;

and (v) Torsion free abelian groups are flat, i.e., if $0 \to B_1 \to B_2 \to B_3 \to 0$ is an exact sequence of abelian groups and if A is a torsion free abelian group then $0 \to A\otimes_Z B_1 \to A\otimes_Z B_2 \to A\otimes_Z B_3 \to 0$ is exact, where Z is the ring of integers.

Let Q be the field of rational numbers and let A be an abelian group. Then $Q\otimes_Z A$ is a Q-vector space. If A is a torsion group then $Q\otimes_Z A = 0$.

If A is a torsion free group define rank(A) $= \dim_Q(Q\otimes_Z A)$. Note that rank(A) = cardinality of a maximal Z-independent subset of A since if $\{a_i\}$ is a maximal Z-independent subset of A then $F = \oplus Za_i$ is a subgroup of A with A/F torsion. Thus, $Q\otimes_Z F \simeq Q\otimes_Z A$ is a Q-vector space with dimension = cardinality of $\{a_i\}$. If rank(A) $= n$ and F is a free subgroup of A of rank n then there is an exact sequence $0 \to F \to Q\otimes_Z A \to T \to 0$ where T is the direct sum of n copies of Q/Z since $0 \to Z \to Q \to Q/Z \to 0$ exact implies that $0 \to Z\otimes_Z F \to Q\otimes_Z F \to (Q/Z)\otimes_Z F \to 0$ is exact. If $0 \to B_1 \to B_2 \to B_3 \to 0$ is an exact sequence of finite rank torsion free abelian groups then $\text{rank}(B_1) + \text{rank}(B_3) = \text{rank}(B_2)$ as a consequence of (v).

An abelian group A is torsion free iff A is isomorphic to a subgroup of $Q\otimes_Z A$ via $a \to 1\otimes a$ since $Z \subseteq Q$ implies that $A \simeq Z\otimes_Z A$ is isomorphic to a subgroup of $Q\otimes_Z A$ whenever A is torsion free by (v). Thus, we may assume that $A \subseteq Q\otimes_Z A$, $(Q\otimes_Z A)/A$ is torsion, and every element of $Q\otimes_Z A$ is of the form qa for $q \in Z$, $a \in A$. Consequently, if B is a subgroup of A with A/B torsion then A/B is isomorphic to a subgroup of the direct sum of rank(A) copies of Q/Z.

If A and B are torsion free of finite rank then $A\otimes_Z B$ is torsion free with $\text{rank}(A\otimes_Z B) = \text{rank}(A)\,\text{rank}(B)$ since a monomorphism $A \to Q\otimes_Z A$ induces a monomorphism $A\otimes_Z B \to (Q\otimes_Z A)\otimes_Z B$ and $(Q\otimes_Z A)\otimes_Z B \simeq (Q\otimes_Z A)\otimes_Z (Q\otimes_Z B)$ is a vector space of dimension = $\dim(Q\otimes_Z A)\,\dim(Q\otimes_Z B)$.

If A and B are torsion free of finite rank and $f \in \text{Hom}_Z(A,B)$ then f extends uniquely to $\bar{f} \in \text{Hom}_Q(Q\otimes_Z A, Q\otimes_Z B)$. Thus $\text{Hom}_Z(A,B)$ is a finite rank torsion free group with rank $\leq \text{rank}(A)\,\text{rank}(B) = \dim_Q(\text{Hom}_Q(Q\otimes_Z A, Q\otimes_Z B))$.

A subgroup B of a torsion free group A is pure in A if $B\cap nA = nB$ for all $n \in Z$. Note that B is pure in A iff A/B is torsion free. If S is a subset of A let $\langle S\rangle$ be the subgroup of A generated by S and $\langle S\rangle_* = \{a \in A \mid na \in \langle S\rangle$ for some $0 \neq n \in Z\}$, the pure subgroup of A generated by S, noting that, in fact, $\langle S\rangle_*$ is a pure subgroup of A and the smallest pure subgroup of A containing S. If $S = \{x\}$ then $\langle S\rangle$ is denoted by Zx and $\langle S\rangle_*$ by $\langle x\rangle_*$.

If B is a pure subgroup of the torsion free group A and if C is an abelian group then $0 \to C\otimes_Z B \to C\otimes_Z A$ is exact. Moreover, if B_i is a pure subgroup of the torsion free group A_i for $i = 1, 2$ then $B_1\otimes_Z B_2$ is isomorphic to a pure subgroup of $A_1\otimes_Z A_2$, the isomorphism being given by $b_1\otimes b_2 \to b_1\otimes b_2$.

If p is a prime of Z and A is an abelian group then $A/pA \simeq (Z/pZ)\otimes_Z A$. Define <u>p-rank(A)</u> $= \dim_{Z/pZ}(A/pA)$. If $0 \to A \to B \to C \to 0$ is an exact sequence of torsion free abelian groups, each having finite p-rank, then p-rank(A) + p-rank(C) = p-rank(B) since $0 \to A/pA \to B/pB \to C/pC \to 0$ is exact.

<u>Theorem 0.1.</u> If A is a torsion free abelian group of finite rank and if $0 \neq n \in Z$ then A/nA is finite. Moreover, p-rank(A) $\leq$ rank(A) for each prime p of Z.

<u>Proof.</u> If $n = p_1^{e_1} \cdots p_h^{e_h}$ is a product of powers of distinct primes of Z then $A/nA \simeq A/p_1^{e_1}A \oplus \cdots \oplus A/p_h^{e_h}A$ since $Z/nZ \simeq Z/p_1^{e_1}Z \oplus \cdots \oplus Z/p_h^{e_h}Z$ and $A/nA \simeq (Z/nZ)\otimes_Z A$. If A/p_iA is finite for each p_i then $A/p_i^{e_i}A$ is finite by induction on e_i.

It is now sufficient to assume that $n = p$ is a prime and to prove that p-rank(A) $= \dim_{Z/pZ}(A/pA) \leq$ rank(A). Let $a_1 + pA, \ldots, a_n + pA$ be independent in A/pA. Then $\{a_1, \ldots, a_n\}$ is a Z-independent subset of A for if $m_1a_1 + \cdots + m_na_n = 0$ with $m_i \in Z$ and g.c.d. $(m_1, \ldots, m_n) = 1$ then $m_ia_i \in pA$ for each i, whence $m_i/p \in Z$ for each i, a contradiction. ///

If T is a torsion abelian group and p is a prime then define $T[p] = \{x \in T \mid px = 0\}$, a Z/pZ-vector space.

<u>Theorem 0.2.</u> Suppose that $0 \to A \to B \to T \to 0$ is an exact sequence of abelian groups where A and B are finite rank torsion free and T is torsion. Then $\dim(T[p])$ + p-rank(B) = p-rank(A) + p-rank(T). In particular, if T is finite then p-rank(A) = p-rank(B).

<u>Proof.</u> There is an exact sequence $0 \to K \to A/pA \to B/pB \to T/pT \to 0$ where $K = (A \cap pB)/pA$. Now $T[p] \simeq C/A$, where $C = \{x \in B \mid px \in A\}$. Define

$\theta : C/A \to K$ by $\theta\ (c + A) = pc + pA$, a well-defined isomorphism. Thus $\dim(K) + \dim(B/pB) = \dim(A/pA) + \dim(T/pT)$ as desired. Finally, if T is finite then $\dim(T[p]) = \dim(T/pT)$ so that $p\text{-rank}(A) = p\text{-rank}(B)$. ///

A torsion free group A is divisible if A is a Q-vector space, i.e., $nA = A$ for all $0 \neq n \in Z$. If A is a torsion free group then there is a unique maximal divisible subgroup $d(A)$ of A with $d(A/d(A)) = 0$. Moreover, $A = d(A) \oplus B$ for some B and B is reduced (i.e., $d(B) = 0$). If A is reduced then the endomorphism ring of A is reduced as a group.

Let p be a prime and define $Z_p = \{m/n \in Q \mid \text{g.c.d.}(n,p) = 1\}$, the localization of Z at p. If A is a group then let $A_p = Z_p \otimes_Z A$, a Z_p-module. If A is finite rank torsion free then $p\text{-rank}(A_p) = p\text{-rank}(A)$ since $A/pA \simeq A_p/pA_p$. Moreover, $A \subseteq A_p \subseteq Q \otimes_Z A$ for each prime p and $A = \cap_p A_p$. If $0 \to A \to B \to C \to 0$ is an exact sequence of abelian groups then $0 \to A_p \to B_p \to C_p \to 0$ is an exact sequence of Z_p-modules. Consequently, if B is torsion free and A is pure in B then B_p is torsion free and A_p is pure in B_p.

EXERCISES

0.1: Prove any statement in Section 0 that you have not previously proved. (Properties of tensor products are standard, e.g., Hungerford [1], and the remaining unproved statements may be found in Fuchs [7].)

§1. Types and rank - 1 groups

A height sequence, $\alpha = (\alpha_p)$, is a sequence of non-negative integers, together with ∞, indexed by the elements of Π, the set of primes of Z. Given a torsion free group A, an element a of A and a prime p of Z, define the p-height of a in A, $h_p^A(a)$, to be n if there is a non-negative integer n with $a \in p^nA \backslash p^{n+1}A$ and ∞ if no such n exists. The height sequence of a in A, $h^A(a) = (h_p^A(a))$, is a height sequence.

If $\alpha = (\alpha_p)$ is a height sequence and m is a positive integer define $m\alpha = (\beta_p)$, where $\beta_p = h_p^Z(m) + \alpha_p$ for each $p \in \Pi$ (agreeing that $\infty + k = \infty$ for each non-negative k). Two height sequences $\alpha = (\alpha_p)$ and $\beta = (\beta_p)$ are equivalent if there are positive integers m and n with $m\alpha = n\beta$, i.e., $\alpha_p = \beta_p$ for all but a finite number of p and $\alpha_p = \beta_p$ if either $\alpha_p = \infty$ or $\beta_p = \infty$. This relation is easily seen to be an equivalence relation. An equivalence class τ of height sequences is called a type, written $\tau = [\alpha]$ for some height sequence α.

Define the type of a in A, $\text{type}_A(a)$, to be $[h^A(a)]$. The group A is homogeneous if any two non-zero elements of A have the same type, the common value being denoted by type(A).

A rank-1 torsion free group A is homogeneous since if a and b are non-zero elements of A then $ma = nb$ for some non-zero integers m and n. Thus, $|m|h^A(a) = h^A(ma) = h^A(nb) = |n|h^A(b)$ so that $\text{type}_A(a) = [h^A(a)] = [h^A(b)] = \text{type}_A(b)$.

For example, $\text{type}(Z) = [(\alpha_p)]$, where $\alpha_p = 0$ for each p and $\text{type}(Q) = [(\beta_p)]$, where $\beta_p = \infty$ for each p. The set of types is a complete set of invariants for torsion free groups of rank 1:

Theorem 1.1.

(a) Suppose A and B are rank-1 torsion free groups. Then A and B are isomorphic iff type(A) = type(B).

(b) If τ is a type then there is a torsion free group A of rank 1 with type(A) = τ.

Proof.

(a) ($\rightarrow$) is a consequence of the observation that if $f : A \rightarrow B$ is an isomorphism then $h^A(a) = h^B(f(a))$ for each $a \in A$.

($\leftarrow$) If $0 \neq a \in A$ and $0 \neq b \in B$ then $[h^A(a)] = [h^B(b)]$. Choose positive integers m and n with $h^A(ma) = mh^A(a) = nh^B(b) = h^B(nb)$. The correspondence $ma \rightarrow nb$ lifts to an isomorphism $f : A \rightarrow B$ since for integers k and ℓ the equation $kx = \ell ma$ has a solution x in A iff $ky = \ell nb$ has a solution y in B. Moreover, the solution of either equation is unique since A and B are torsion free.

(b) Let $\tau = [(\alpha_p)]$ and define A to be the subgroup of Q generated by $\{1/p^i | p \in \Pi, 0 \leq i \leq \alpha_p\}$. Then $\text{type}(a) = \tau$, since $h^A(1) = (\alpha_p)$. ///

The set of height sequences has a partial ordering given by $\alpha = (\alpha_p) \leq \beta = (\beta_p)$ if $\alpha_p \leq \beta_p$ for each $p \in \pi$. The operations $\underline{\sup\{\alpha, \beta\}} = (\max\{\alpha_p, \beta_p\})$ and $\underline{\inf\{\alpha, \beta\}} = (\min\{\alpha_p, \beta_p\})$ induce a lattice structure on the set of height sequences. The set of height sequences is closed under the operation $\underline{\alpha + \beta} = (\alpha_p + \beta_p)$.

If $a, b \in A$, a torsion free group, then $h^A(a+b) \geq \inf\{h^A(a), h^A(b)\}$. Furthermore, if $A = A_1 \oplus A_2$ and $a_i \in A_i$ then $h^A(a_1+a_2) = \inf\{h^A(a_1), h^A(a_2)\}$.

The partial order on the set of height sequences induces a partial order on the set of types where $\underline{\tau \leq \sigma}$ if there is $\alpha \in \tau$ and $\beta \in \sigma$ with $\alpha \leq \beta$. To show, for example, that this relation is anti-symmetric assume that $\tau \leq \sigma$ and $\sigma \leq \tau$. There are $\alpha, \alpha' \in \tau$ and $\beta, \beta' \in \sigma$ with $\alpha \leq \beta$ and $\beta' \leq \alpha'$. Choose positive integers m and n with $m\beta = n\beta'$. Then $m\alpha \leq m\beta = n\beta' \leq n\alpha'$ so that $km\beta = n\alpha'$ for some positive integer k. Thus $\sigma = [\beta] = [\alpha'] = \tau$. For types $\tau = [\alpha]$ and $\sigma = [\beta]$, define $\underline{\tau + \sigma} = [\alpha+\beta]$, a well-defined operation.

The partial order on the set of types corresponds to the existence of non-zero homomorphisms between rank-1 groups:

Proposition 1.2. Let A and B be rank-1 torsion free groups. Then the following are equivalent:

(a) $\mathrm{Hom}(A,B) \neq 0$;

(b) There is a monomorphism $A \to B$;

(c) $\mathrm{type}(A) \leq \mathrm{type}(B)$.

Proof. (a) $\to$ (b) Since A and B are rank-1 torsion free every non-zero homomorphism from A to B is a monomorphism.

(b) $\to$ (c) If $f : A \to B$ is a monomorphism and $0 \neq a \in A$ then $0 \neq f(a) \in B$ with $h^A(a) \leq h^B(f(a))$.

(c) $\to$ (a) Choose $0 \neq a \in A$ and $0 \neq b \in B$ with $h^A(a) \leq h^B(b)$. Then $a \to b$ extends to a non-zero homomorphism $A \to B$. ///

Corollary 1.3. Let A and B be torsion free groups of rank 1. Then the following are equivalent:

(a) A and B are isomorphic;

(b) $\mathrm{Hom}(A,B) \neq 0$ and $\mathrm{Hom}(B,A) \neq 0$;

(c) There is a monomorphism $f : A \to B$ such that $B/f(A)$ is finite.

Proof. (a) $\to$ (c) is clear.

(c) $\to$ (b) Suppose that $nB \leq f(A) \leq B$ for some $0 \neq n \in Z$. Then $0 \neq f^{-1}n : B \to A$ is a well defined homomorphism.

(b) $\to$ (a) As a consequence of Corollary 1.2, $\mathrm{type}(A) = \mathrm{type}(B)$ so that Theorem 1.1 applies. ///

The following theorem gives a description of quotients of rank-1 torsion free groups, as well as an alternative definition for the height of an element. For a prime p and a non-negative integer i, let $Z(p^i)$ denote the cyclic group of order p^i. Define $Z(p^\infty)$ to be the p-torsion subgroup of Q/Z. If C is a subgroup of $Z(p^\infty)$ then $C \simeq Z(p^i)$ for some $0 \leq i < \infty$ and $Z(p^\infty)/C \simeq Z(p^\infty)$, or $C = Z(p^\infty)$.

Theorem 1.4: Assume that $0 \neq b \in B$, a subgroup of a rank-1 torsion free group A. Then $A/B \simeq \oplus_p Z(p^{i_p})$ where $h^A(b) = (\ell_p)$, $h^B(b) = (k_p)$, and $i_p = \ell_p - k_p$ (agreeing that $\infty - \infty = 0$ and $\infty - k = \infty$ if $k < \infty$).

Proof: In view of the exact sequence $0 \to B/Zb \to A/Zb \to A/B \to 0$ it is sufficient to prove that $A/Zb \simeq \oplus_p Z(p^{\ell_p})$. As noted in Section 0, A/Zb is isomorphic to a subgroup of Q/Z so that $A/Zb \simeq \oplus_p C_p$ with $C_p \subseteq Z(p^\infty)$. But $C_p \simeq Z(p^{\ell_p})$ since $p^i x = nb$ with g.c.d. $(p,n) = 1$ has a solution $x \in A$ iff $i \leq h_p^A(b) = h_p^A(nb) = \ell_p$. ///

The types of $\mathrm{Hom}(A,B)$ and $A \otimes_Z B$, where A and B are rank-1 groups, may be computed from the types of A and B. A type τ is non-nil if there is $\alpha = (\alpha_p) \in \tau$ with $\alpha_p = 0$ or ∞ for each p.

Theorem 1.5: Suppose that A and B are torsion free groups of rank-1.

(a) If $\mathrm{type}(A) \leq \mathrm{type}(B)$ then $\mathrm{Hom}(A,B)$ is a rank-1 torsion free group with type $= [(m_p)]$, where $0 \neq a \in A$, $0 \neq b \in B$, $h^A(a) = (k_p) \leq h^B(b) = (\ell_p)$, $m_p = \infty$ if $\ell_p = \infty$, and $m_p = \ell_p - k_p$ if $\ell_p < \infty$.

(b) If $\mathrm{type}(A) = [(k_p)]$ then $\mathrm{type}(\mathrm{Hom}(A,A)) = [(m_p)]$ is non-nil, where $m_p = \infty$ if $k_p = \infty$ and $m_p = 0$ if $k_p < \infty$.

(c) $A \otimes_Z B$ is a torsion free group of rank 1 with $\mathrm{type}(A \otimes_Z B) = \mathrm{type}(A) + \mathrm{type}(B)$.

Proof:

(a) As noted in Section 0, the group $\mathrm{Hom}(A,B)$ is torsion free of rank 1. Define $\phi : \mathrm{Hom}(A,B) \to B$ by $\phi(f) = f(a)$. Then $\mathrm{Image}(\phi) \subseteq G = \{x \in B \mid h^B(x) \geq h^A(a)\}$ since $h^B(f(a)) \geq h^A(a)$ for each $f \in \mathrm{Hom}(A,B)$. On the other hand, $G \subseteq \mathrm{Image}(\phi)$ for if $x \in B$ and $h^B(x) \geq h^A(a)$ then there is $f : A \to B$ with $f(a) = x$. Hence $\mathrm{type}(\mathrm{Hom}(A,B)) = \mathrm{type}(G) = [(m_p)]$ as desired.

(b) is a consequence of (a).

(c) There is an embedding $A\otimes_Z B \to Q\otimes_Z Q \simeq Q$ so that $A\otimes_Z B$ is torsion free of rank 1. If $0 \neq a \in A$ and $0 \neq b \in B$ then $h^{A\otimes B}(a\otimes b) \geq h^A(a) + h^B(b)$ since $a = ma'$ and $b = nb'$ imply that $a\otimes b = mn(a'\otimes b')$. Therefore, $\text{type}(A\otimes_Z B) \geq \text{type}(A) + \text{type}(B)$.

To show that $\text{type}(A\otimes_Z B) \leq \text{type}(A) + \text{type}(B)$, it suffices to prove that if $p^i x = a\otimes b$, with $h_p^A(a) = h_p^B(b) = 0$, has a solution x in $A\otimes_Z B$ then $i = 0$. By Exercise 1.4, $x = a'\otimes b'$ for some $a' \in A$, $b' \in B$. Thus $p^i(a'\otimes b') = a\otimes b$. Choose non-zero integers k and ℓ with $ka' = \ell a$ and p prime to k. Then $k(a\otimes b) = (p^i ka')\otimes b' = (p^i \ell a)\otimes b'$ so that $a\otimes(kb) = a\otimes(p^i \ell b')$. Thus $kb = p^i \ell b'$, hence $i = 0$, since g.c.d. $(p,k) = 1$ and $h_p^B(b) = 0$. ///

Corollary 1.6. Assume that A is a torsion free group of rank-1. The following are equivalent:

(a) A has non-nil type;

(b) $\text{type}(A) + \text{type}(A) = \text{type}(A)$;

(c) $A \simeq \text{Hom}(A,A)$;

(d) A is isomorphic to the additive group of a subring of Q;

(e) $A\otimes_Z A \simeq A$;

(f) If $0 \neq a \in A$ then $A/Za = T \oplus D$, where T is a finite torsion group and $D \simeq \oplus_{p\in S} Z(p^\infty)$ for some subset S of Π.

Proof: Exercise 1.9. ///

The preceding results, due essentially to Baer [2], demonstrate that the set of types is a useful set of invariants for torsion-free groups of rank 1. In general, however, the type of a group provides little information about the structure of the group.

The remainder of this section is devoted to several generalizations of the notion of type, due to Warfield [1] and Richman [1].

For a torsion free group A let $\text{typeset}(A) = \{\text{type}_A(a) \mid 0 \neq a \in A\}$. Note that $\text{type}_A(a) = \text{type}(\langle a\rangle_*)$, where $\langle a\rangle_*$ is the pure rank-1 subgroup of A generated by $0 \neq a \in A$. The group A is homogeneous iff $\text{typeset}(A)$ has cardinality 1.

Assume that A is a torsion free group of rank n, $S = \{x_1, x_2, \ldots, x_n\}$ is a maximal Z-independent subset of A, and let $X_i = \langle x_i\rangle_*$. Then $X = X_1 \oplus \ldots \oplus X_n \subseteq A$ and A/X is a torsion group. Define the <u>inner type of A</u>, $IT(A)$, to be $\inf\{\text{type}(X_1), \ldots, \text{type}(X_n)\}$.

<u>Proposition 1.7.</u>

(a) $IT(A)$ is independent of S.

(b) $IT(A) \leq \tau$ for each $\tau \in \text{typeset}(A)$.

(c) If $\text{typeset}(A)$ is finite then $IT(A) \in \text{typeset}(A)$.

<u>Proof</u>:

(a) Let $S' = \{x_1', \ldots, x_n'\}$ be another maximal Z-independent subset of A, $X_i' = \langle x_i'\rangle_*$, $X' = X_1' \oplus \ldots \oplus X_n' \subseteq A$, and $\tau' = \inf\{\text{type}(X_1') \ldots, \text{type}(X_n')\}$. Since A/X is torsion, for each i there is $0 \neq m_i \in Z$ and $k_j \in Z$ with $m_i x_i' = k_1 x_1 + \ldots + k_n x_n$. Thus $\text{type}(X_i') = \text{type}_A(x_i') = \text{type}_A(m_i x_i') \geq \inf\{\text{type}_A(k_j x_j)\} \geq \inf\{\text{type}_A(x_j)\} = IT(A)$, and so $\tau' \geq IT(A)$. Similarly, $IT(A) \geq \tau'$.

(b) Let $\tau = \text{type}_A(x) \in \text{typeset}(A)$, $0 \neq x \in A$. Extend x to a maximal Z-independent subset of A. As a consequence of (a), $\tau \geq IT(A)$.

(c) If typeset(A) is finite then there is a pure rank-1 subgroup X of A such that whenever $\tau \in \text{typeset}(A)$ and $\tau \leq \text{type}(X)$ then $\tau = \text{type}(X)$. It is sufficient to prove that if Y is another pure rank-1 subgroup of A then $\text{type}(X) \leq \text{type}(Y)$; in which case $\text{type}(X) = IT(A)$.

Let $0 \neq x \in X$ and $0 \neq y \in Y$. Then $\{x+ny \mid n \in Z\}$ is infinite and $\text{typeset}(A)$ is finite so there are non-zero distinct integers m and n with $\text{type}_A(y+nx) = \text{type}_A(y+mx)$. But

$\text{type}_A(x) = \text{type}_A((n-m)x) = \text{type}_A((y+nx) - (y+mx)) \geq \text{type}_A(y+mx) = \text{type}_A(y+nx)$.

By the choice of X, $\text{type}_A(y+nx) = \text{type}_A(x) = \text{type}_A(nx)$. Finally, $\text{type}(Y) = \text{type}_A(y) = \text{type}_A((y+nx) - nx) \geq \text{type}_A(nx) = \text{type}(X)$, as desired. ///

There is a rank-2 group A with $IT(A) \notin \text{typeset}(A)$ (Example 2.8). Let A be a torsion free group of rank n and let $S = \{x_1, \ldots, x_n\}$ be a maximal Z-independent subset of A. Define $Y_i = \langle x_1, \ldots, x_{i-1}, x_{i+1}, \ldots, x_n \rangle_*$ and define the outer type of A, $OT(A)$, to be $\sup \{\text{type}(A/Y_1), \ldots, \text{type}(A/Y_n)\}$ noting that each A/Y_i is torsion free of rank-1.

Proposition 1.8. Suppose that A is a torsion free group of finite rank.

(a) $OT(A)$ is independent of S.

(b) If A/B is a torsion free rank-1 quotient of A then $\text{type}(A/B) \leq OT(A)$.

Proof.

(a) follows from (b).

(b) Let X be a rank-1 group with $OT(A) = \text{type}(X)$ (Theorem 1.1.b) There is a homomorphism $f : A \to A/Y_1 \oplus \ldots \oplus A/Y_n$, defined by $f(a) = (a+Y_1, \ldots, a+Y_n)$. Since $\text{type}(A/Y_i) \leq \text{type}(X)$, there is a monomorphism $A/Y_i \to X$ (Corollary 1.2). Combining the preceding homomorphisms gives a monomorphism $g : A \to X^n$, the direct sum of n copies of X. Let C be the pure subgroup of X^n generated by $g(B)$. Then $A/B \to X^n/C$, given by $a + B \to g(a) + C$ is a monomorphism since $\text{rank}(A/B) = 1$.

It is now sufficient to prove that $X^n/C \simeq X$; in which case Corollary 1.2 implies that $\text{type}(A/B) \leq \text{type}(X)$. As a consequence of Exercise 1.2 $X^n = C \oplus C'$ and $C' \simeq X^m$ for some m. But $n-1 = \text{rank}(B) = \text{rank}(C)$ so $\text{rank}(C') = 1$ and $X \simeq C' \simeq X^n/C$. ///

If rank $A = 1$ then it is clear that $\text{type}(A) = IT(A) = OT(A)$. Two torsion groups T_1 and T_2 are quasi-isomorphic if there are homomorphisms $f : T_1 \to T_2$ and $g : T_2 \to T_1$ such that $T_2/f(T_1)$ and $T_1/g(T_2)$ are bounded. Let A be a finite rank torsion free group and $S = \{x_1, x_2, \ldots, x_n\}$ a maximal Z-independent subset of A. Then $F = Zx_1 \oplus \ldots \oplus Zx_n$ is a free subgroup of A with A/F torsion. Define the Richman type of A, $RT(A)$, to be the equivalence class of A/F under quasi-isomorphism of torsion groups.

Proposition 1.9: Let A be a torsion free group of finite rank,

(a) $RT(A)$ is independent of F.

(b) Suppose that $\operatorname{rank}(A) = 1$. Then $RT(A)$ is the equivalence class of $\oplus_p Z(p^{i_p})$ where $\operatorname{type}(A) = [(i_p)]$.

Proof.

(a) Let F_1 and F_2 be free subgroups of A with A/F_i torsion, $i = 1, 2$. Since $F_1 \cap F_2$ is free with $A/(F_1 \cap F_2)$ torsion it suffices to assume that $F_1 \subseteq F_2$. There is an exact sequence $0 \to F_2/F_1 \to A/F_1 \to A/F_2 \to 0$ and F_2/F_1 is finite, since both F_1 and F_2 are free groups with the same rank. By Exercise 1.10, A/F_1 and A/F_2 are quasi-isomorphic torsion groups, as needed.

(b) follows from Theorem 1.4. ///

The following proposition shows how $IT(A)$ and $OT(A)$ may be computed from $RT(A)$ with Proposition 1.9.b as a special case, due to Warfield [1].

Theorem 1.10: Assume that A is a torsion free group of rank n and that F is a free subgroup of A with A/F torsion. Write $A/F = \oplus_p T_p$, where $T_p = Z(p^{i_{p,1}}) \oplus \ldots \oplus Z(p^{i_{p,n}})$ with $0 \le i_{p,1} \le i_{p,2} \le \ldots \le i_{p,n} \le \infty$. Then $IT(A) = [(i_{p,1})]$ and $OT(A) = [(i_{p,n})]$.

Proof. To show that $OT(A) = [(i_{p,n})]$, let $F = Zx_1 \oplus \ldots \oplus Zx_n$ and $Y_i = \langle x_1, \ldots, x_{i-1}, x_{i+1}, \ldots, x_n\rangle_*$. There is an exact sequence $0 \to Z(x_i + Y_i) \to A/Y_i \to A/(Zx_i \oplus Y_i) \to 0$ so that $A/(Zx_i \oplus Y_i) \simeq \oplus_p Z(p^{k_{p,i}})$ with $\operatorname{type}(A/Y_i) = [(k_{p,i})]$ (Theorem 1.4). Therefore, for each p and each $1 \le i \le n$ $i_{p,n} = \max\{i_{p,j}\} \ge k_{p,i}$, since there is an epimorphism

$A/F \to A/(Zx_i \oplus Y_i) \to 0$ given by $a + F \to a + (Zx_i \oplus Y_i)$. Thus, $OT(A) = \sup \{type(A/Y_i)\} \leq [(i_{p,n})]$. On the other hand, there is a monomorphism $A/F \to A/(Zx_1 \oplus Y_1) \oplus \ldots \oplus A/(Zx_n \oplus Y_n)$ given by $a + F \to (a + (Zx_1 \oplus Y_1), \ldots, a + (Zx_n \oplus Y_n))$ since $\cap\{Zx_i \oplus Y_i | 1 \leq i \leq n\} = F$. Hence $i_{p,n} = \max \{i_{p,j}\} \leq \max \{k_{p,j}\}$ and $[(i_{p,n})] \leq OT(A) = \sup \{type(A/Y_j)\}$.

Finally, to show that $IT(A) = [(i_{p,1})]$ let $X_i = \langle x_i \rangle_*$ and $X_i/Zx_i = \oplus_p Z(p^{\ell_{p,i}})$ so that $type(X_i) = [(\ell_{p,i})]$ (Theorem 1.4). There is an exact sequence $0 \to (X_1/Zx_1) \oplus \ldots \oplus (X_n/Zx_n) \to A/F \to A/(X_1 \oplus \ldots \oplus X_n) \to 0$. Thus $m_p = \min \{\ell_{p,j}\} \leq i_{p,1} = \min \{i_{p,j}\}$ for each p and $IT(A) = \inf \{type(X_i)\} = [(m_p)] \leq [(i_{p,1})]$. Conversely, A/F may be embedded in $(Q/Z)^n$ so that $\dim(A/F)[p] \leq n$ for each p. If $\dim(A/F)[p] < n$ then $i_{p,1} = 0$ and $m_p = i_{p,1}$. Otherwise, $\dim(A/F)[p] = n$. There is an exact sequence $0 \to (F+X_j)/X_j \to A/X_j \to A/(F+X_j) \to 0$ so that $\dim(A/(F+X_j))[p] \leq n-1$, since $rank(A/X_j) = n-1$. In view of the exact sequence $0 \to X_j/Zx_j \to A/F \to A/(F+X_j) \to 0$ there is some k with $\ell_{p,j} = i_{p,k}$. Thus, $m_p = \min \{\ell_{p,j}\} \geq \min \{i_{p,j}\} = i_{p,1}$ so that $IT(A) = [(m_p)] \geq [(i_{p,1})]$ as needed. ///

While the invariants $IT(A)$, $OT(A)$, and $RT(A)$ need not completely describe the group A, in certain cases they provide some information about the structure of A and associated homomorphism groups.

Theorem 1.11 (Warfield [1]). Suppose that A and B are torsion free groups of rank m and n, respectively. Then $rank(Hom(A,B)) \leq mn$. Moreover, $rank(Hom(A,B)) = mn$ iff $IT(B) \geq OT(A)$.

Proof. As mentioned in Section 0 $rank(Hom(A,B)) \leq mn$. Furthermore, $rank(Hom(A,B)) = mn$ iff for each $g : Q\otimes A \to Q\otimes B$ there is $0 \neq k \in Z$ with $kg : A \to B$, i.e., $Hom(Q\otimes A, Q\otimes B)/Hom(A,B)$, is torsion.

Assume that $\mathrm{Hom}(A,B)$ has rank mn, and let C be a pure rank-1 subgroup of B and A/D a rank-1 torsion free quotient of A. The exact sequence $0 \to D \to A \to A/D \to 0$ induces an exact sequence $0 \to Q\otimes D \to Q\otimes A \to Q\otimes(A/D) \to 0$. Now $Q\otimes(A/D) \simeq Q\otimes C \simeq Q$ so the composite homomorphism $g : Q\otimes A \to Q\otimes(A/D) \to Q\otimes C \to Q\otimes B$ is in $\mathrm{Hom}(Q\otimes A, Q\otimes B)$. Choose $0 \neq k \in Z$ with $kg : A \to B$. Then $kg(A) \subseteq (\mathrm{Image}\ (Q\otimes C)) \cap B = C$ since C is pure in B and $\mathrm{Kernel}(kg) = \mathrm{Kernel}(g) \cap A = (Q\otimes D) \cap A = D$. Thus kg induces $0 \neq f : A/D \to C$. Therefore, $\mathrm{type}(A/D) \leq \mathrm{type}(C)$ so that $OT(A) \leq IT(B)$.

Conversely, assume that $IT(B) \geq OT(A)$. Let $B_1, \ldots, B_n$ be pure rank-1 subgroups of B with $B_1 \oplus \ldots \oplus B_n \subseteq B$, $IT(B) = \inf \{\mathrm{type}(B_i)\}$ and let $A/A_1, \ldots, A/A_m$ be rank-1 quotients of A with $OT(A) = \sup \{\mathrm{type}(A/A_j)\}$. Since $\mathrm{type}(A/A_j) \leq OT(A) \leq IT(B) \leq \mathrm{type}(B_i)$ there is a non-zero monomorphism $f_{ji} : A/A_j \to B_i$ for each $1 \leq j \leq m$ and $1 \leq i \leq n$. It follows that $\{f_{ji}\}$ induces a Z-independent subset of $\mathrm{Hom}(A,B)$ with cardinality mn. ///

The next lemma is commonly referred to as Baer's Lemma. The classical proof is given below. A proof in a more general setting is given in Section 5.

Lemma 1.12. Suppose that A is finite rank torsion free and that A/B is rank-1 torsion free. If $a = x + b$ with $b \in B$ and $\mathrm{type}_A(x) \geq \mathrm{type}(A/B)$ for each $a \in A$ then B is a summand of A.

Proof. It is sufficient to find $x \in A \setminus B$ with $h^A(x) = h^C(x+B)$, where $C = A/B$. Given such an x there is $g : C \to A$ with $g(x+B) = x$. Let $\Pi : A \to C$ be the canonical homomorphism. Then $\Pi g - 1_C$ is an endomorphism of the rank-1 group C with non-zero kernel so that $\Pi g = 1_C$ and $A = g(C)\oplus B$ as needed.

Choose $a \in A \setminus B$ with $\text{type}_A(a) \geq \text{type}(C)$. Then $h^A(a) \leq h^C(a+B)$. Now $\text{type}_A(a) = \text{type}_C(a+B)$ there is $0 \neq m \in Z$ with $h^A(ma) = h^C(a+B)$. Write $a+B = m'(a'+B)$ where $h_p^C(a'+B) = 0$ for each prime p dividing m. Choose $0 \neq n \in Z$ with $h^A(na') = h^C(a'+B)$, as above. Then $\text{g.c.d.}(n,m) = 1$, say $rn + sm = 1$ for some $r, s \in Z$.

Define $x = rnm'a' + sma$. Then $x = rnm'a' + sma = rnm'a' + a - rna$ and $x = m'a' - smm'a' + sma$. Thus $x + B = a + B = m'a' + B$, since $rn(m'a'-a) \in B$ and $sm(-m'a'+a) \in B$. Finally,
$h^A(x) \leq h^C(x+B) = h^C(a+B) = h^C(m'a'+B) = \inf\{h^A(ma), h^A(nm'a')\} \leq \inf\{h^A(sma), h^A(rnm'a')\} \leq h^A(rnm'a'+sma) = h^A(x)$. Thus $h^A(x) = h^C(x+B)$. ///

Corollary 1.13 (Warfield [1]. Let A be a finite rank torsion free group. Then the following are equivalent:

(a) $\text{rank}(\text{Hom}(A,A)) = (\text{rank } A)^2$;

(b) $IT(A) = OT(A)$;

(c) $A \simeq C^n$ for some $0 < n \in Z$ and a rank-1 group C.

Proof.

(a) $\leftrightarrow$ (b) is Theorem 1.11 noting that $IT(A) \leq OT(A)$.

(b) $\rightarrow$ (c) Let A/B be a torsion free rank-1 quotient of A. If $a \in A \setminus B$ then $IT(A) \leq \text{type}_A(a) \leq \text{type}_{A/B}(a+B) = \text{type}(A/B) \leq OT(A)$ so that $\text{type}_A(a) = \text{type}(A/B)$. By Baer's Lemma, $A = B \oplus C$ for some $C \simeq A/B$. But $IT(A) \leq IT(B) \leq OT(B) \leq OT(A)$ so that $IT(B) = OT(B)$. By induction on $n = \text{rank}(A)$, $B \simeq C^{n-1}$, noting that $\text{type}(C) = \text{type}(A/B) = IT(B) = OT(B)$. Thus, $A \simeq C^n$.

(c) $\rightarrow$ (a) If $A \simeq C^n$ with $\text{rank } C = 1$ then $n = \text{rank } A$. Moreover, $\text{Hom}(C^n, C^n) \simeq (\text{Hom}(C,C))^{n^2}$ has rank n^2. ///

Two special cases for $RT(A)$ can occur: (i) $RT(A) = [\oplus_p T_p]$ where $|T_p| < \infty$ for all p with $pA \neq A$; (ii) $RT(A) = [\oplus_p T_p]$ where

$pT_p = T_p$ for each p. These two cases give rise to two special classes of groups.

For a finite rank torsion free group A let R(A) be the subring of Q generated by $\{1/p \mid p \in \Pi,\ pA = A\}$. Define A to be R(A) - locally free if for each $p \in \Pi$ $p\text{-rank}(A) = 0$ or else $p\text{-rank}(A) = \text{rank}(A)$.

Theorem 1.14. The following are equivalent for a finite rank torsion free group A:

(a) A is $R(A)$-locally free;

(b) If A/B is a rank-1 torsion free quotient of A and if p is a prime then $pA = A$ iff $p(A/B) = A/B$;

(c) $OT(A) = [(m_p)]$, where each $m_p < \infty$ for each prime p with $pA \neq A$;

(d) $RT(A) = [\oplus_p T_p]$, where $|T_p| < \infty$ for each prime p with $pA \neq A$.

Proof.

(a) $\to$ (b) If $pA = A$ then $p(A/B) = A/B$. Conversely, assume that $p(A/B) = A/B$. Tensoring the pure exact sequence $0 \to B \to A \to A/B \to 0$ by Z/pZ gives an exact sequence $0 \to B/pB \to A/pA \to (A/B)/p(A/B) \to 0$. Thus, $p\text{-rank}(A) = p\text{-rank}(B) \le \text{rank}(B) < \text{rank}(A)$. Since A is $R(A)$-locally free, $p\text{-rank}(A) = 0$, i.e., $pA = A$.

(b) $\to$ (c) follows from the fact that $OT(A) = \sup\{\text{type}(A/A_i)\}$, where each A/A_i is a torsion free rank-1 quotient of A.

(c) $\to$ (d) is a consequence of Theorem 1.10.

(d) $\to$ (a) There is an exact sequence $0 \to F \to A \to T \to 0$ where F is a free subgroup of A, $T = \oplus_p T_p$ is torsion, and each T_p is finite whenever $pA \neq A$. By Theorem 0.2, $p\text{-rank}(A) + \dim(T[p]) = p\text{-rank}(F) + \dim(T/pT)$. But $\dim(T/pT) = \dim(T[p])$, whenever $pA \neq A$, in which case $p\text{-rank}(A) = p\text{-rank}(F) = \text{rank}(F) = \text{rank}(A)$. ///

<u>Theorem 1.15</u>. Assume that A is a rank-1 torsion free group and that B is a finite rank torsion free group. Then $B \simeq \text{Hom}(G,A)$ for some finite rank torsion free G iff $R(B) = R(A)$ and $\text{type}(A) \geq OT(B)$. In this case, B is $R(B)$-locally free and G may be chosen to be $\text{Hom}(B,A)$.

<u>Proof.</u> ($\rightarrow$) If $pA = A$ then $pB = B$ since $f \in \text{Hom}(G,A)$ implies that $(1/p)f \in \text{Hom}(G,A)$. Thus $R(A) \subseteq R(B)$. Let $0 \rightarrow F \rightarrow G \rightarrow T \rightarrow 0$ be an exact sequence with F free and T torsion. Then $0 \rightarrow 0 = \text{Hom}(T,A) \rightarrow \text{Hom}(G,A) \rightarrow \text{Hom}(F,A) \simeq A^n$ is exact, where $n = \text{rank}(G)$. Thus there is a monomorphism $B \rightarrow A^n$ so that $OT(B) = OT(\text{Hom}(G,A)) \leq \text{type}(A)$ (Exercise 1.1). Finally to prove that $R(B) \subseteq R(A)$, assume that $pB = B$. Then $pA = A$ since $OT(B) \leq \text{type}(A)$.

($\leftarrow$) Note that B is $R(B)$-locally free since if $OT(B) = [(m_p)] \leq \text{type}(A) = [(k_p)]$ with $m_p \leq k_p$, $R(A) = R(B)$, and $pB \neq B$ then $pA \neq A$ and $m_p \leq k_p < \infty$. Thus, B is $R(B)$-locally free by Theorem 1.14.

It is sufficient to prove that $\phi : B \rightarrow \text{Hom}(\text{Hom}(B,A),A)$, defined by $\phi(b)(f) = f(b)$, is an isomorphism. Now ϕ is monic, since if $\phi(b) = 0$ then $b \in \cap\{\text{Kernel}(f) \mid f \in \text{Hom}(B,A)\} = 0$ (Exercise 1.1, since $\text{type}(A) \geq OT(B)$).

By Theorem 1.11, $\text{rank}(\text{Hom}(B,A)) = \text{rank}(B)\ \text{rank}(A) = \text{rank}(B)$. As in the proof of ($\rightarrow$), $\text{type}(A) \geq OT(\text{Hom}(B,A))$. Similarly, $\text{rank}(\text{Hom}(\text{Hom}(B,A),A)) = \text{rank}(\text{Hom}(B,A)) = \text{rank}(B)$. Since ϕ is a monomorphism, it now suffices to prove that $\text{Image}(\phi)$ is pure.

Suppose that $\theta : \text{Hom}(B,A) \rightarrow A$ and $p\theta = \phi(b)$ for some prime p. If $b \in pB$ then $\theta \in \text{Image}(\phi)$. Otherwise, $b \in B \backslash pB$ so that $pA \neq A$ since $R(A) = R(B)$. Let $\{b_1+pB, \ldots, b_n+pB\}$ be a basis of B/pB with $b = b_1$. Then $F = Zb_1 \oplus \ldots \oplus Zb_n \subseteq B$ with $(B/F)_p = 0$. Let $F' = Zb_z \oplus \ldots \oplus Zb_n$ and let X be the pure subgroup of B generated by F'. Then $((B/X)/(Zb+X))_p = 0$. Since $OT(B) \leq \text{type}(A)$ then is $a \in A \backslash pA$ with $h^{B/X}(b+X) \leq h^A(a)$ and an embedding $h : B/X \rightarrow A$ with $h(b+X) = a$. Then $hf : B \rightarrow A$ where $f : B \rightarrow B/X$ and $hf(b) = \phi(b)(hf) \notin pA$, a contradiction.///

Define a finite rank torsion free group A to be quotient divisible if A has a free subgroup F such that $A/F = T \oplus D$ where T is finite and D is torsion divisible; equivalently, $RT(A) = [\oplus_p T_p]$ where $pT_p = T_p$ for each p. The next corollary characterizes the groups A that are both $R(A)$-locally free and quotient divisible.

Corollary 1.16. A finite rank torsion free group A is quotient divisible and $R(A)$-locally free iff $A \simeq B^n$ for some $0 < n \in Z$, where B is a rank-1 group with non-nil type. In this case, $B \simeq R(A)$.

Proof. ($\leftarrow$) Since type(B) is non-nil there is $0 \neq b \neq B$ with $B/Zb = \oplus_p Z(p^{i_p})$ where $i_p = 0$ if $pB \neq B$ and $i_p = \infty$ if $pB = B$. Thus, A has a free subgroup F with $A/F = \oplus_p T_p$ where $T_p = 0$ if $pA \neq A$ and T_p is divisible if $pA = A$. Therefore A is quotient divisible, since A/F is torsion divisible, and $R(A)$-locally free by Theorem 1.14.

($\rightarrow$) It suffices to assume that A has a free subgroup F with A/F divisible and torsion, since if $A/F' = T \oplus D$, where T is finite and D is torsion divisible, then $A/F = D$, where $F = \{a \in A \mid a + F' \in T\}$ is free (if $nT = 0$ then $nF' \subseteq F \subseteq F'$). As a consequence of Theorem 1.14, $T = A/F = \oplus_p T_p$, where $T_p = 0$ if $pA \neq A$. The exact sequence $0 \rightarrow F \rightarrow A \rightarrow T \rightarrow 0$ induces an exact sequence $0 \rightarrow R(A) \otimes F \rightarrow R(A) \otimes A \rightarrow R(A) \otimes T \rightarrow 0$. But $R(A) \otimes F \simeq R(A)^n$, where $n = \text{rank}(A)$, and $R(A) \otimes A \simeq A$. Moreover, $R(A) \otimes T = 0$, since if $pA = A$ then $R(A)$ is p-divisible and T_p is torsion. Thus $A \simeq R(A)^n$ and $R(A)$ is a rank-1 group with non-nil type (Corollary 1.6). ///

Quotient divisible groups are discussed in detail by Beaumont-Pierce [1].

EXERCISES

1.1 (Warfield [1]): Let A and B be torsion free groups with rank(A) = 1 and rank(B) = n.

(a) Prove that the following are equivalent:

(i) type(A) $\geq$ OT(B);

(ii) rank(Hom(B,A)) = n;

(iii) There is a monomorphism $f : B \to A^n$;

(iv) $\cap \{\text{Kernel}(f) \mid f \in \text{Hom}(B,A)\} = 0$

(b) Prove that the following are equivalent:

(i) type(A) $\leq$ IT(B);

(ii) rank(Hom(A,B)) = n;

(iii) There is a monomorphism $f : A^n \to B$;

(iv) The subgroup of B generated by $\{g(A) \mid g \in \text{Hom}(A,B)\}$ is equal to B.

1.2 (J. Reid): Suppose that A and G are finite rank torsion free groups and that rank(A) = 1.

(a) Prove that there is a 1-1 correspondence from pure subgroups of G to pure subgroups of $A \otimes_Z G$ given by $C \to A \otimes_Z C$.

(b) Prove that if $B \simeq A^n$ and if C is a pure subgroup of B then C is a summand of B and $C \simeq A^m$ for some m (Hint: use (a) and the fact that $A^n \simeq A \otimes_Z (Z^n)$).

1.3 (a) Let A and B be torsion free groups of rank-1. Prove that $A \otimes_Z A \simeq B \otimes_Z B$ iff $A \simeq B$.

(b) Let C be a rank-1 torsion free group. Prove that type(C) = $[(m_p)]$ with $m_p < \infty$ for each p iff whenever A and B are rank-1 torsion free groups with $A \otimes_Z C \simeq B \otimes_Z C$ then $A \simeq B$.

1.4 Suppose that A and B are torsion free groups with rank(A) = 1. Show that if $x \in A \otimes_Z B$ then $x = a \otimes b$ for some $a \in A$, $b \in B$. (Hint: consider the case that $1 \in A \subseteq Q$).

1.5 Suppose that A is a torsion free group of finite rank and that $nA \subseteq B \subseteq A$ for some $0 \neq n \in Z$ where $B = A_1 \oplus \ldots \oplus A_k$ and each A_i is a pure rank-1 subgroup of A. Prove that A/B is generated, as a group, by $\leq k-1$ elements (Hint: Consider the local case of the last paragraph of the proof of Theorem 1.10).

1.6 Show that if $A_1, A_2, \ldots, A_n$ are subgroups of Q then:

(a) $\text{type}(A_1 + \ldots + A_n) = \sup \{\text{type}(A_i) | 1 \leq i \leq n\}$,

(b) $\text{type}(\cap \{A_i | 1 \leq i \leq n\}) = \inf \{\text{type}(A_i) | 1 \leq i \leq n\}$.

1.7 Suppose that A is a torsion free group of finite rank and that $nA \subseteq B \subseteq A$ for some $0 \neq n \in Z$ where $B = A_1 \oplus \ldots \oplus A_k$ and each A_i is a rank-1 subgroup of A. Let $k^+ = \{1, 2, \ldots, k\}$. Prove each of the following statements:

(a) $\text{typeset}(A) = \{\inf \{\text{type}(A_i) | i \in I\} | I$ is a non-empty subset of $k^+\}$.

(b) $\text{IT}(A) = \inf \{\text{type}(A_i) | 1 \leq i \leq k\}$.

(c) $\text{OT}(A) = \sup \{\text{type}(A_i) | 1 \leq i \leq k\}$.

(d) $\text{RT}(A)$ is the equivalence class of $A_1/Za_1 \oplus \ldots \oplus A_k/Za_k$ where each $0 \neq a_i \in A_i$.

(e) p-rank(A) is the cardinality of $\{i | 1 \leq i \leq k$ and $pA_i \neq A_i\}$.

(f) A is $R(A)$-locally free iff whenever p is a prime with $pA \neq A$ then $pA_i \neq A_i$ for each i.

(g) A is quotient divisible iff each A_i has non-nil type.

(h) $A \otimes_Z A \simeq A$ iff $k = 1$ and A_1 has non-nil type.

1.8 Let A be a torsion free group of finite rank. For a prime p let $A_p = Z_p \otimes_Z A$ where Z_p is the localization of Z at p. Prove that A is $R(A)$-locally free iff A_p is a free Z_p-module for each prime p with $pA \neq A$.

1.9 Prove Corollary 1.6.

1.10 Suppose that T_1 and T_2 are subgroups of finite direct sums of copies of Q/Z. Show that T_1 and T_2 are quasi-isomorphic iff there is a homomorphism $f : T_1 \to T_2$ with Kernel(f) and $T_2/f(T_1)$ finite.

§2. Examples of Indecomposable Groups and Direct Sums

A finite rank torsion free group A is completely decomposable if A is the direct sum of rank-1 groups and almost completely decomposable if there is $0 \neq k \in Z$ and a completely decomposable subgroup B of A with $kA \subseteq B \subseteq A$. If $kA \subseteq A_1 \oplus \ldots \oplus A_n \subseteq A$ with $\operatorname{rank}(A_i) = 1$ for each i then $kA \subseteq \langle A_1\rangle_* \oplus \ldots \oplus \langle A_n\rangle_* \subseteq A$. Consequently, A is almost completely decomposable iff there is $0 \neq k \in Z$ and pure rank-1 subgroups $A_1, \ldots, A_n$ of A with $kA \subseteq A_1 \oplus \ldots \oplus A_n \subseteq A$.

Completely decomposable groups are well behaved with respect to direct sum decompositions. On the other hand, as illustrated below, almost completely decomposable groups may have non-equivalent direct sum decompositions into indecomposable groups.

Many of the following examples are classical and, as such, the constructions are sketched. The reader is referred to Fuchs [7], Vol. II, for details, as well as for numerous other examples and references.

Lemma 2.1: Suppose that A is finite rank torsion free and that $nA \subseteq B \oplus C \subseteq A$ for some $0 \neq n \in Z$. If X is a fully invariant subgroup of A then $nX \subseteq (B \cap X) \oplus (C \cap X) \subseteq X$.

Proof: Write $n = \sigma_B + \sigma_C$ where σ_B and σ_C are quasi-projections (multiplication by n followed by a projection of $B \oplus C$ onto B and C respectively). Then $nX = \sigma_B(X) + \sigma_C(X) \subseteq (B \cap X) \oplus (C \cap X) \subseteq X$. ///

Example 2.2: For each integer $n > 1$ there is an indecomposable almost completely decomposable group of rank n.

Proof: Choose subgroups $A_1, A_2, \ldots, A_n$ of Q such that $\operatorname{type}(A_i)$ and $\operatorname{type}(A_j)$ are incomparable, i.e. $\operatorname{Hom}(A_i, A_j) = 0$, whenever $i \neq j$; $1 \in A_i$ for each i; and there is a prime p with $h_p^{A_i}(1) = 0$ for each i. Let $V = Qx_1 \oplus \ldots \oplus Qx_n$ and define A to be the subgroup of V generated by $\{a_i x_i \mid a_i \in A_i,\ 1 \le i \le n\} \cup \{(x_1 + \ldots + x_n)/p\}$. The group A is denoted by $\langle A_1x_1, \ldots, A_nx_n, (x_1 + \ldots + x_n)/p\rangle$.

Then (i) A is almost completely decomposable since $pA \subseteq A_1x_1 \oplus \ldots \oplus A_nx_n$; (ii) each A_ix_i is pure in A so that $\text{type}_A(x_i) = \text{type}(A_i)$ (iii) each A_ix_i is fully invariant in A since $\text{Hom}(A_i, A_j) = 0$ if $i \ne j$; and (iv) A is indecomposable as a consequence of (ii), (iii), and the fact that each A_i is indecomposable (recall that if $a = b + c \in B \oplus C = A$ then $h_p^A(a) = \inf\{h_p^B(b), h_p^C(c)\}$ and note that $A/(A_1x_1 \oplus \ldots \oplus A_nx_n) \simeq Z/pZ$). ///

The construction of Example 2.2 requires some restriction on the types of the A_i's to guarantee that A is indecomposable, as illustrated by a theorem of Beaumont-Pierce [2].

Theorem 2.3: Suppose that A is finite rank torsion free with rank-1 subgroups $A_1, \ldots, A_n$ such that $kA \subseteq A_1 \oplus \ldots \oplus A_n \oplus C \subseteq A$ for some $0 \ne k \in Z$. If $\text{type}(A_1) \le \text{type}(A_2) \le \ldots \le \text{type}(A_n) \le IT(C)$ then $A \simeq A_1 \oplus A_2 \oplus \ldots \oplus A_n \oplus C$.

Proof: Let $B = \langle A_2, A_3, \ldots, A_n, C\rangle_*$ so that A/B is torsion free of rank 1 and $k(A/B) \subseteq (A_1 + B)/B \subseteq A/B$. Then $\text{type}(A/B) = \text{type}(A_1)$, since $(A_1 + B)/B \simeq A_1$ and A/B has rank 1 (Corollary 1.3).

Now $A = B \oplus D$ for some $D \simeq A/B \simeq A_1$, as a consequence of Baer's Lemma (Lemma 1.12): if $a \in A\backslash B$ then $ka = a_1 + \ldots + a_n + c$ for some $a_i \in A_i$, $c \in C$, and $a_1 \ne 0$. Thus, $\text{type}_A(a) = \text{type}_A(ka) = \inf\{\text{type}_A(a_i)\} = \text{type}(A_1) = \text{type}(A/B)$.

Finally, $kB \subseteq A_2 \oplus \ldots \oplus A_n \oplus C \subseteq B$ so by induction on rank(A), $A \simeq A_1 \oplus A_2 \oplus \ldots \oplus A_n \oplus C$. ///

A slight modification of the construction in Example 2.2 gives rise to a group with considerably different properties.

A torsion free group A is strongly indecomposable if whenever $0 \ne k \in Z$ and $kA \subseteq B \oplus C \subseteq A$ then $B = 0$ or $C = 0$. Note that strongly indecomposable groups are indecomposable, but not conversely (Example 2.2). Also, A is strongly indecomposable iff whenever $0 \neq k \in Z$ with $kA \subseteq B \oplus C \subseteq A$ and B, C pure subgroups of A then $B = 0$ or $C = 0$.

Example 2.4: For each integer $n > 1$ there is a strongly indecomposable torsion free abelian group A of rank n with $\text{rank}(\text{Hom}(A,A)) = 1$.

Proof: Choose subgroups $A_1, \ldots, A_n$ of Q such that $\text{type}(A_i)$ and $\text{type}(A_j)$ are incomparable whenever $i \neq j$; $1 \in A_i$ for each i and there is a prime p with $h_p^{A_i}(1) = 0$ for each i.

Let $V = Qx_1 \oplus \ldots \oplus Qx_n$ and define A to be the subgroup of V generated by $A_1x_1, \ldots, A_nx_n$ and $\{(x_1 + \ldots + x_n)/p^j \mid j = 1, 2, \ldots\}$. The group A is denoted by $\langle A_1x_1, \ldots, A_nx_n, (x_1 + \ldots + x_n)/p^\infty\rangle$.

Then (i) A_ix_i is pure in A and $\text{type}_A(x_i) = \text{type}(A_i)$; (ii) A_ix_i is fully invariant in A; and (iii) A is strongly indecomposable as a consequence of (ii), Lemma 2.1, the fact that each A_i is indecomposable, and the fact that $A/(A_1x_1 \oplus \ldots \oplus A_nx_n) \simeq Z(p^\infty)$.

Note that $\text{type}_A(x_1 + \ldots + x_n) = [(m_q)]$ where $\text{type}(A_i) = [(m_{q,i})]$, $m_q = \min\{m_{q,i}\}$ if $q \neq p$ and $m_p = \infty$. Thus, $X = \langle x_1 + \ldots + x_n\rangle_*$ is pure and fully invariant in A.

If $f : A \to A$ then $f(x_i) = q_ix_i$ for some $q_i : A_i \to A_i$ and $f(x_1 + \ldots + x_n) = q(x_1 + \ldots + x_n)$ for some $q : X \to X$. Thus $q = q_i$ for each i so that f is multiplication by $q \in Q$, i.e., there is an embedding $\text{Hom}(A,A) \to Q$. ///

Let $E(A)$ denote the endomorphism ring of A. The group A, constructed in Example 2.2, has $E(A)$ isomorphic to a subring of finite index in $E(A_1) \times \ldots \times E(A_n)$. The duality developed in Section 5 together with Example 2.11 show that finitely generated projective $E(A)$-modules need not be free in this case.

On the other hand, the group A constructed in Example 2.4 has $E(A)$ isomorphic to a subring of Q so that $E(A)$ is a principal ideal domain (Exercise 2.2).

The duality in Section 5 is used to show that, in this case, direct sum decompositions of A^m are equivalent.

In contrast to Theorem 2.3, there are strongly indecomposable groups A with pure rank 1 subgroups $A_1, \ldots, A_n$ such that $A/(A_1 \oplus \ldots \oplus A_n)$ is torsion and $\text{type}(A_1) \leq \text{type}(A_2) \leq \ldots \leq \text{type}(A_n)$.
If τ_1 and τ_2 are types define $\tau_1 << \tau_2$ if there is a type τ with $\tau_1 = [(k_p)] < \tau = [(\ell_p)] < \tau_2 = [(m_p)]$, $k_p \leq \ell_p \leq m_p$ for each p, and $k_p = \ell_p$ whenever $m_p = \infty$. Then $\tau_1 << \tau_2$ iff $\{p \mid k_p < m_p < \infty\}$ is infinite. For example, if $\tau_1 = [(0, 0, \ldots)]$ and $\tau_2 = [(\infty, \infty, 0, 0, \ldots)]$ then it is <u>not</u> true that $\tau_1 << \tau_2$.

<u>Example 2.5</u>: For each integer $n \geq 1$ and types $\tau_1 << \tau_2 << \ldots << \tau_n$ there is a strongly indecomposable group A of rank n with pure rank 1 subgroups $A_1, \ldots, A_n$ such that $A/(A_1 \oplus \ldots \oplus A_n)$ is torsion and $\text{type}(A_i) = \tau_i$ for each i.

<u>Proof</u>: The construction is by induction on n. The case $n = 1$ is clear, so assume that $n > 1$. Choose a strongly indecomposable group B of rank n-1 with pure rank-1 subgroups $A_2, \ldots, A_n$ such that $\text{type}(A_i) = \tau_i$ for $i \geq 2$ and $B/(A_2 \oplus \ldots \oplus A_n)$ is torsion. Since $\tau_1 << \tau_2$ there is a rank-1 group C with $\tau_1 = [(k_p)] < \text{type}(C) = [(\ell_p)] < \tau_2 = [(m_p)]$, $k_p \leq \ell_p \leq m_p$ for each p and $k_p = \ell_p$ whenever $m_p = \infty$. Thus, there is an exact sequence $0 \to Z(p^{k_p}) \to Z(p^{\ell_p}) \to Z(p^{\ell_p - k_p}) \to 0$, where $\infty - \infty = 0$ and $\infty - k = \infty$ if $k < \infty$. Moreover, there is an embedding $Z(p^{\ell_p - k_p}) \to T_p$ for each p where $(Q \otimes_Z B)/B = \oplus_p T_p$ and each T_p is a p-torsion group (noting that $T_p = 0$ iff $m_p = \infty$ since $\tau_2 = IT(B)$). Choose $c_0 \in C$ with $C/Zc_0 \simeq \oplus_p Z(p^{\ell_p})$. Then there is $f : C \to (Q \otimes_Z B)/B$ with $Zc_0 \subseteq \text{Kernel}(f)$ and $\text{Kernel}(f)/Zc_0 \simeq \oplus_p Z(p^{k_p})$ induced by the preceeding exact sequences and embeddings , one for each p.

Let A be the pullback of $f : C \to (Q \otimes_Z B)/B$ and the natural map $\pi : Q \otimes_Z B \to (Q \otimes_Z B)/B$, i.e., $A = \{(x,c) \mid x \in Q \otimes_Z B,\ c \in C, \text{ and } f(c) = x + B\}$.

Then there is a commutative diagram with exact rows:

$$\begin{array}{ccccccccc} 0 & \to & B & \xrightarrow{i} & A & \xrightarrow{h} & C & \to & 0 \\ & & \downarrow 1_B & & \downarrow g & & \downarrow f & & \\ 0 & \to & B & \xrightarrow[j]{} & Q\otimes_Z B & \xrightarrow[\pi]{} & (Q\otimes_Z B)/B & \to & 0 \end{array}$$

where $i(b) = (b,0)$, $g(x,c) = x$, $h(x,c) = c$, and $j(b) = 1 \otimes b$.

Moreover, there is $0 \neq a \in A$ such that $A_1/Za \simeq \text{Kernel}(f)/Zc_0$ where $A_1 = \text{Kernel}(g) = \langle a\rangle_*$, since $h : \text{Kernel}(g) \to \text{Kernel}(f)$ is an isomorphism. Consequently, $\text{type}(A_1) = \tau_1 = [(k_p)]$. (Theorem 1.4).

Note that B is fully invariant in A for if $\alpha : A \to A$ then the composite $B \xrightarrow{\alpha} A \to A/B$ is zero since $\tau_2 = IT(B) > \text{type}(C) = \text{type}(A/B)$, i.e. $\alpha(B) \subseteq B$. Furthermore, each A_i is a pure rank-1 subgroup of A with $\text{type}(A_i) = \tau_i$ and $A/(A_1 \oplus \ldots \oplus A_n)$ is torsion.

To show that A is strongly indecomposable, assume that $0 \neq k \in Z$ and that $kA \subseteq X \oplus Y \subseteq A$ with X,Y non-zero pure subgroups of A. Since B is fully invariant in A, $kB \subseteq (X\cap B) \oplus (Y\cap B) \subseteq B$ (Lemma 2.1). But B is strongly indecomposable so $X \cap B = 0$ or $Y \cap B = 0$, say $X \cap B = 0$. Then $kB \subseteq Y \cap B \subseteq Y$ and $B \subseteq Y$ since Y is pure in A. A computation of ranks shows that $B = Y$, i.e., $kA \subseteq X \oplus B \subseteq A$. But $kC \subseteq (X \oplus B)/B \subseteq C = A/B$, hence $\text{type}(X) = \text{type}(C)$. On the other hand, $ka = x + b$ for some $x \in X$, $b \in B$. Thus $\text{type}(C) > \tau_1 = \text{type}_A(ka) = \inf\{\text{type}_X(x), \text{type}_B(b)\} \geq \inf\{\text{type}(C), IT(B)\} = \text{type}(C)$, a contradiction. ///

Remark: It is proved in a later section that if A is strongly indecomposable then every endomorphism of A is a monomorphism or else nilpotent. The group A, constructed in Example 2.5, is a strongly indecomposable group with a non-zero nilpotent endomorphism since there is a non-zero homomorphism $C \to B$, $(\text{type}(C) < \tau_2 = IT(B))$, which induces a non-zero homomorphism $A \to C = A/B \to B$ that is not a monomorphism.

A similar construction gives rise to a slightly different example which is used in a later section to describe the structure of a certain Grothendieck group due to Rotman [2].

Example 2.6: If A is a rank-1 torsion free group then there is a torsion free group G of rank 2 with exact sequences $0 \to A \to G \to Q \to 0$ and $0 \to E(A) \to G \to Q \to 0$. Furthermore, $IT(G) = \text{type}(E(A))$ and $OT(G) = \text{type}(Q)$.

Proof: Let $\text{type}(A) = [(k_p)]$ so that $\text{type}(E(A)) = [(m_p)]$ where $m_p = \infty$ if $k_p = \infty$ and $m_p = 0$ if $k_p < \infty$ (Theorem 1.5.b). Thus $(Q \otimes A)/A \simeq \oplus_p \{Z(p^\infty) \mid pA \neq A\} \simeq (Q \otimes E(A))/E(A)$. Choose an epimorphism $f : Q \to (Q \otimes A)/A$ with $\text{Kernel}(f) \simeq E(A)$. If G is the pullback of f and $Q \otimes A \to (Q \otimes A)/A$ there is a commutative diagram with exact rows

$$\begin{array}{ccccccccc} 0 & \longrightarrow & A & \longrightarrow & G & \longrightarrow & Q & \longrightarrow & 0 \\ & & \downarrow{\scriptstyle 1_A} & & \downarrow{\scriptstyle g} & & \downarrow{\scriptstyle f} & & \\ 0 & \longrightarrow & A & \longrightarrow & Q \otimes A & \longrightarrow & (Q \otimes A)/A & \longrightarrow & 0 \end{array}$$

where $\text{Kernel}(g) \simeq \text{Kernel}(f) \simeq E(A)$. Since f is onto $Q \simeq Q \otimes A = g(G) + A$ and $\text{type}(Q) = \sup\{\text{type}(g(G)), \text{type}(A)\}$. Also, since $A \cap \text{Kernel}(g) = 0$, $A \simeq g(A) \subseteq g(G)$ hence $\text{type}(A) \leq \text{type } g(G)$. Consequently, $\text{type}(g(G)) = \text{type}(Q)$ and $0 \to \text{Kernel}(g) \to G \to Q \to 0$ is exact with $\text{Kernel}(g) \simeq E(A)$. Finally, $IT(G) = \inf\{\text{type}(A), \text{type}(\text{Kernel}(g))\} = \text{type}(E(A))$ and $OT(G) = \sup\{\text{type}(G/A), \text{type}(G/E(A))\} = \text{type}(Q)$. ///

As a consequence of Theorem 2.3, homogeneous almost completely decomposable groups must be completely decomposable. The following example shows, among other things, that there are homogeneous strongly indecomposable torsion free groups of arbitrarily large finite rank.

Example 2.7: Given $1 \leq r \in Z$ and $1 < n \in Z$ there is a homogeneous strongly indecomposable torsion free group A of rank r such that $\text{type}(A) = \text{type}(Z)$, $E(A) \simeq Z$, $A/nA \simeq (Z/nZ)^r$ and if B is a subgroup of A with rank $B \leq r-1$ then B is free.

Proof: Let p be a prime not dividing n, choose $\sigma_2, \ldots, \sigma_r \in Z_p^* \setminus pZ_p^*$ algebraically independent over Q, where Z_p^* is the ring of p-adic integers, and let $\sigma_1 = 1$.

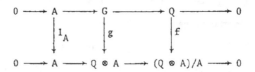

Represent each σ_j as an infinite sum $\Sigma s_{ij}p^i$ where each s_{ij} is an integer with $0 \le s_{ij} < p$. For each $2 \le j \le r$ let $\sigma_{ij} = s_{0j} + s_{1j}p + \ldots + s_{i-1,j}p^{i-1} \in Z$ and define $\sigma_{i1} = 1 = \sigma_1$. Note that as $i \to \infty$, $\sigma_{ij} \to \sigma_j$.

Define $A = \langle x_1, \ldots, x_r, y_1, y_2, \ldots \rangle \subseteq Qx_1 \oplus \ldots \oplus Qx_r$ where $y_i = (x_1 + \sigma_{i2}x_2 + \ldots + \sigma_{ir}x_r)/p^i$ for $i = 1, 2, \ldots$. Then $\text{rank}(A) = r$. If $q \ne p$ is a prime then $A_q = Z_q \otimes A \simeq Z_q x_1 \oplus \ldots \oplus Z_q x_r$, since $\sigma_{ij}/p^i \in Z_q$. Thus $A/qA \simeq A_q/qA_q \simeq (Z_q/qZ_q)^r \simeq (Z/qZ)^r$. Since p does not divide n, it follows that $A/nA \simeq (Z/nZ)^r$.

For each i, $py_{i+1} = y_i + s_{i2}x_2 + \ldots + s_{ir}x_r$. Thus, $A/F \simeq Z(p^\infty)$ where $F = Zx_1 \oplus \ldots \oplus Zx_r$. Let B be a pure subgroup of A with $\text{rank}(B) \le r - 1$. Then $B/(F \cap B) \to A/F$, given by $b + (F \cap B) \to b + F$, is an embedding with Image $= (B+F)/F$. If $B/(F \cap B)$ is finite then $B \simeq F \cap B$ is free. Otherwise, $(B+F)/F = A/F$ and $A = B + F$. Now $0 \to B \to A \to A/B \to 0$ is exact and $A/B \simeq F/(F \cap B)$ is free. Thus $A = B \oplus F'$ for some free group F'.

It is now sufficient to prove that $E(A) \simeq Z$; in which case A is strongly indecomposable and the preceding paragraph shows that every subgroup B of rank $\le r-1$ is free, noting that $B \subseteq \langle B \rangle_*$ is pure with $\text{rank}(\langle B \rangle_*) = \text{rank}(B)$.

If $a \in A$ then $a = k_2x_2 + \ldots + k_rx_r + ky_s$ for some k_i, k, $s \in Z$, since $py_{i+1} = y_i + s_{i2}x_2 + \ldots + s_{ir}x_r$. Moreover, if $a \in A \backslash F$ then s and k may be chosen with g.c.d. $(k, p) = 1$. Consequently, $C = Zx_2 \oplus \ldots \oplus Zx_r$ is p-pure, hence pure in A. Thus, if $p^m a \in F$ then $p^m a = p^m(k_2x_2 + \ldots + k_rx_r) + kp^m y_s = n_2x_2 + \ldots + n_rx_r + k'y_m$ with $p^m | k'$. Since C is pure in A, $p^m a = p^m(n_2'x_2 + \ldots + n_r'x_r) + k'y_m$ for some n_i', $k' \in Z$.

If $f \in E(A)$ then there is $0 \ne t \in Z$ with $tf : F \to F$. Thus, it suffices to assume $f : F \to F$ say $f(x_i) = \Sigma m_{ij}x_j$. For each k,

$p^k f(y_k) = \Sigma_i\ \sigma_{ki} f(x_i) = \Sigma_i\ \Sigma_j\ \sigma_{ki}\ m_{ij}\ x_j = p^k(n_{k2}\ x_2 + \ldots + n_{kr}\ x_r) + n_k y_k \in F$ where n_{kj}, $n_k \in Z$. Equating coefficients of the x_j's gives $\Sigma_i\ \sigma_{ki}\ m_{i1} = n_k/p^k \in Z$ and $\Sigma_i\ \sigma_{ki}\ m_{ij} = p^k n_{kj} + n_k(\sigma_{kj}/p^k) = p^k n_{kj} + (\Sigma_i\ \sigma_{ki}\ m_{i1})\ \sigma_{kj}$ if $j \geq 2$. Letting $k \to \infty$ gives $\Sigma_i\ \sigma_i\ m_{ij} = (\Sigma_i\ \sigma_i\ m_{i1})\ \sigma_j$ if $j \geq 2$. Since $\sigma_1 = 1$ and $\{\sigma_2, \ldots, \sigma_r\}$ are algebraically independent over Q, $m_{jj} = m_{11}$ for $2 \leq j \leq r$ and $m_{ij} = 0$ if $i \neq j$. Thus, f is multiplication by m_{11} and $E(A) \simeq Z$. ///

Example 2.8: For each integer $n > 1$ there is a strongly indecomposable group A of rank n such that typeset(A) is infinite and $IT(A) \notin$ typeset(A). Moreover, if B is a pure subgroup of A then B is strongly indecomposable and $E(B) \simeq Z$.

Proof. Let $V = Qx_1 \oplus \ldots \oplus Qx_n$ and $T = \{r_1x_1 + \ldots + r_nx_n | r_i \in Z$ with g.c.d. $(r_1, \ldots, r_n) = 1\}$. Enumerate $T = \{t_1, t_2, \ldots\}$ and write $\Pi = \cup S_i$ as the disjoint union of infinitely many infinite sets. Define A to be the subgroup of V generated by $\{(1/p_i)t_i | t_i \in T, p_i \in S_i, i = 1, 2, \ldots\}$.

The reader may verify that if $0 \neq a \in A$ then there are relatively prime integers k and ℓ with $ka = \ell t_i$ for some $t_i \in T$ (c.g., if $a = \Sigma\ m_i\ (1/p_i)t_i$ then $p_1p_2 \ldots p_n\ a = m\ (\Sigma\ n_ix_i)$ with g.c.d. $(n_1, \ldots, n_n) = 1$). Thus typeset(A) = $\{\text{type}_A(t_i) | i = 1, 2, \ldots\}$. Furthermore, $\text{type}_A(t_i) = [(m_{p,i})]$ where $m_{p,i} = 1$ if $p \in S_i$ and $m_{p,i} = 0$ if $p \in \Pi\backslash S_i$ (equate coefficients of the x_i's and use the fact that $\cup S_i$ is a disjoint union).

Note that $IT(A) = \text{type}(Z) \notin$ typeset(A) since if $0 \neq a \in A$ then $\text{type}_A(a) = \text{type}_A(t_j)$ for some j and $\text{type}(Z) = \inf\{\text{type}_A(t_i), \text{type}_A(t_j)\}$ whenever $i \neq j$.

Let B be a pure subgroup of A and assume that $kB \subseteq C \oplus D \subseteq B$ for some $0 \neq k \in Z$ with $0 \neq D$ and $0 \neq C$. Write $b = c + d$ with

$0 \neq c \in C$ and $0 \neq d \in D$. Then $\text{type}_A(b) = \text{type}_B(b) = \inf\{\text{type}_A(c), \text{type}_A(d)\} = \text{type}(Z)$, recalling that B is pure in A, a contradiction.

To see that $E(B) \simeq Z$, let $b_1, \ldots, b_m$ be a maximal Z-independent subset of B. After a possible relabelling of T, $B_i = \langle b_i \rangle_* = \langle t_i \rangle_*$ for $1 \leq i \leq n$. If $f \in E(B)$ then $f : B_i \to B_i$, a pure fully invariant subgroup of A and B, by the construction of A. But $E(B_i) \simeq Z$ (Theorem 1.5) so $f(b_i) = q_i b_i$ for some $q_i \in Z$. Moreover, $\langle b_1 + \ldots + b_m \rangle_* = B_j = \langle t_j \rangle_* \subseteq B$ for some $j > m$ so that $f : B_j \to B_j$ and $f(b_1 + \ldots + b_m) = q(b_1 + \ldots + b_m)$ for some $q \in Z$. Thus f is multiplication by $q \in Z$. ///

The group A in Example 2.8 has the property that any two distinct pure rank-1 subgroups have incomparable types. The proof of Example 2.8 shows that if any two pure rank-1 subgroups of A have incomparable type then each pure subgroup B of A is strongly indecomposable and E(B) is isomorphic to a subring of Q, hence a principal ideal domain. Torsion free groups with the property that every pure subgroup is indecomposable are called purely indecomposable and are considered by Griffith [1] and Armstrong [2].

The next series of examples illustrates the pathological behavior of direct sum decompositions of finite rank torsion free groups. The first three examples are due to B. Jónsson [1] and [2]. Generalizations of these constructions are given in Fuchs [7], Vol. II, pp. 134-140.

Example 2.9: There are indecomposable groups A, B, C, D of ranks 1, 3, 2, 2, respectively, with $A \oplus B = C \oplus D$.

Proof: Let $V = Qx \oplus Qy \oplus Qz \oplus Qu$ and let $\langle x/p^\infty \rangle$ denote the group Cx where $1 \in C \subset Q$ with $h_p^C(1) = \infty$ and $h_q^C(1) = 0$ if $q \neq p$.

Let $x' = 3x - y$, $y' = 2x - y$ so that $V = Qx' \oplus Qy' \oplus Qz \oplus Qu$ and $x = x' - y'$, $y = 2x' - 3y'$.

Define $A = \langle x/5^\infty \rangle$, $B = \langle y/5^\infty, z/7^\infty, u/11^\infty, (y+z)/3, (y+u)/2 \rangle$, $C = \langle x'/5^\infty, z/7^\infty, (x'-z)/3 \rangle$, and $D = \langle y'/5^\infty, u/11^\infty, (y'-u)/2 \rangle$. Then A, B, C, and D are indecomposable (as in Example 2.2) and the reader may verify that $A \oplus B = C \oplus D$. ///

Example 2.10: There are indecomposable groups A, B, C, and D of ranks 1, 2, 1, 2, respectively, such that $A \oplus B = C \oplus D$, $A \simeq C$, and $B \not\simeq D$.

Proof: Let $V = Qx \oplus Qy \oplus Qz$, $x' = 8x + 3y$, $y' = 5x + 2y$ so that $V = Qx' \oplus Qy' \oplus Qz$, $x = 2x' = 3y'$, and $y = 5x' + 8y'$. Let P_1 and P_2 be two infinite disjoint sets of primes with $5 \notin P_1 \cup P_2$; $R_1 = \{n \in Z | n$ is square free with all prime divisors in $P_1\}$ and $R_2 = \{m \in Z | m$ is square free with all prime divisors in $P_2\}$.

Define $B = \langle y/r, z/s, (y+z)/5 | r \in R_1, s \in R_2 \rangle$, $A = \langle x/r | r \in R \rangle$, $C = \langle x'/r | r \in R_1 \rangle$ and $D = \langle y'/r, z/s, (3y'+z)/5 | r \in R_1, s \in R_2 \rangle$. Then A, B, C, D are indecomposable; $A \oplus B = C \oplus D$; $A \simeq C$; but B and D are not isomorphic. ///

Example 2.11: There are indecomposable rank 2 groups A and C such that $A \oplus A \simeq C \oplus C$ yet A and C are not isomorphic.

Proof: Let $V = Qx \oplus Qy \oplus Qz \oplus Qu$, $x' = 2x + z$, $y' = y + 3u$, $z' = 17x + 9z$, and $u' = y + 2u$. Then $V = Qx' \oplus Qy' \oplus Qz' \oplus Qu'$ and $x = 9x' - z'$, $y = 2y' + 3u'$, $z = 17x' + 2z'$, and $u = y' - u'$. Let P_1, P_2, R_1 and R_2 be as defined in the previous example.

Define
$$A = \langle x/r, u/s, (x+y)/5 | r \in R_1, s \in R_2 \rangle$$
$$B = \langle z/r, y'/s, (z+u)/5 | r \in R_1, s \in R_2 \rangle$$
$$C = \langle x'/r, y'/s, (x' + 2y')/5 | r \in R_1, s \in R_2 \rangle$$
$$D = \langle z'/r, u'/s, (z' + 2u')/5 | r \in R_1, s \in R_2 \rangle.$$

Then $A \simeq B$, $C \simeq D$ and $A \oplus B = C \oplus D$. Moreover, as in the preceeding example, A and C are not isomorphic. ///

Two direct sum decompositions $A = A_1 \oplus \ldots \oplus A_m = B_1 \oplus \ldots \oplus B_n$ of A into indecomposable summands are equivalent if $m = n$ and there is a permutation σ of $\{1, 2, \ldots, n\}$ with $A_i \simeq B_{\sigma(i)}$ for each i.

Example 2.12: For each integer $n \geq 1$ there is a rank 3 torsion free group A which has at least n non-equivalent direct sum decompositions.

Proof: Fuchs [7], Vol. II, Theorem 90.4. ///

It is proved in Theorem 11.11 that if A is a finite rank torsion free group then A has, up to isomorphism, only finitely many summands. Thus, A has only finitely many non-equivalent direct sum decompositions.

Examples 2.9 - 2.12 suggest that the endomorphism ring of a finite rank torsion free group can be complicated. In fact, the following theorem, known as Corner's Theorem, shows that essentially every subring of a finite dimensional Q-algebra may be realized as the endomorphism ring of a finite rank torsion free group.

Theorem 2.13: Suppose that R is a ring and that the additive group of R is reduced torsion free of rank $n < \infty$. Then there is a reduced torsion free group A of rank 2n such that $R \simeq E(A)$.

Proof: Fuchs [7], Vol. II, Theorem 110.2. ///

Corner [2] gives an example of a ring of rank n that cannot be realized as the endomorphism ring of a group with rank < 2n. In certain cases, a ring of rank n can be realized as the endomorphism ring of a group of rank n.

Theorem 2.14: Let R be a ring such that R^+, the additive group of R, is torsion free of finite rank n.

(a) (Zassenhaus [1]) If R^+ is a free abelian group then $R \simeq E(A)$ for some torsion free group A of rank n.

(b) (Butler [2]) If for each prime p, $R_p = Z_p \otimes_Z R$ is a free Z_p-module then $R \simeq E(A)$ for some torsion free group of rank n.

If p is a prime then a group A is <u>p-local</u> if $qA = A$ for each prime $q \neq p$; equivalently A is a Z_p-module.

The construction given in Examples 2.9-2.12 are not applicable to p-local groups. In fact, it is proved in Corollary 7.19 that if A, B, C, are finite rank p-local torsion free groups then (i) $A \oplus B \simeq A \oplus C$ implies that $B \simeq C$ and (ii) $A \oplus A \simeq B \oplus B$ implies that $A \simeq B$. The following example is a variation of an unpublished example of M.C.R. Butler.

Example 2.15: There is a finite rank torsion free p-local group G with non-equivalent direct sum decompositions.

Proof: Let $p \geq 5$ be a prime and choose a Z_p-algebra R such that R is an integral domain having at least four distinct principal maximal ideals $M_i = f_iR$; e.g., $h(x) = x^4 + 6x^3 + (11 + p)x^2 + 6x + 2p = x(x+1)(x+2)(x+3) + p(x^2+2)$ is irreducible in $Q[x]$, hence $Z_p[x]$, by Eisenstein's criterion. Let $R = Z_p[\alpha]$ for some root α of $h(x)$ so that $M_1 = R\alpha$, $M_2 = R(\alpha+1)$, $M_3 = R(\alpha+2)$, and $M_4 = R(\alpha+3)$ are distinct maximal ideals of R.

Choose a p-local finite rank torsion free group A with $R = E(A)$ (Theorem 2.13). Since R is a domain, A must be indecomposable.

There are ideals I_1 and I_2 of R and subgroups A_1 and A_2 of A with $E(A) = I_1 + I_2$, $pA \subseteq I_iA \subseteq A_i \subset A$ and $A \neq A_i$ for $i = 1,2$. E.g. let $I_1 = M_1M_2$ and $I_2 = M_3M_4$ so that $E(A) = I_1 + I_2$ and $M_iA = f_iA \neq A$ for each i. Choose $a_1 \in M_2A \setminus M_1A$, $a_2 \in M_1A \setminus M_2A$ so that $a = a_1 + a_2 \in A \setminus (M_1A \cup M_2A)$.

Define $A_i = I_iA + Za$. Then $pA \subseteq I_iA \subseteq A_i$ since $pE(A) \subseteq I_i$. Assume that $A_1 \simeq A$, say $f(A) = A_1$ for some $f \in E(A)$. Then $I_1A \subseteq A_1 = f(A)$, $f^{-1}I_1(A) \subseteq A$, $f^{-1}I_1 \subseteq E(A)$, and $I_1 \subseteq fE(A)$. But $f \notin M_1 \cup M_2$, since $a \in f(A) \setminus (M_1A \cup M_2A)$, so that $E(A)f = E(A)$ (noting that M_1 and M_2 are the only maximal ideals containing I_1). Therefore, $A_1 = A$ and $Z/pZ \simeq A_1/I_1A = A/I_1A \simeq A/M_1A \oplus A/M_2A$, a contradiction. Similarly, $A_2 \neq A$.

Write $1_A = f_1 + f_2$ with $f_i \in I_i$ and define $\theta : A \to A_1 \oplus A_2$ by $\theta(a) = (f_1(a), f_2(a))$ and $\phi : A_1 \oplus A_2 \to A$ by $\phi(a_1,a_2) = a_1 + a_2$. Then $\phi\theta = 1_A$ so that $G = A \oplus \text{Kernel } \theta = A_1 \oplus A_2$. Therefore, G has at least two non-equivalent direct sum decompositions since A is indecomposable, $\text{rank}(A) = \text{rank}(A_i)$ and $A \neq A_i$ for $i = 1,2$.

EXERCISES

2.1 Let A and X be torsion free groups of finite rank with rank(X) = 1 and let $G = X \otimes_Z A$.

(a) Prove that typeset(G) = $\{\text{type}(X) + \tau \mid \tau \in \text{typeset}(A)\}$

(b) Prove that if G is strongly indecomposable then A is strongly indecomposable.

(c) Give an example showing that if A is strongly indecomposable then G need not be strongly indecomposable.

2.2 Prove that every subring of Q is a principal ideal domain.

2.3 (Wickless) Let $n > 0$ be a square free integer and $Q(\sqrt{n}) = \{a + b\sqrt{n} \mid a, b \in Q\}$, a quadratic number field. Define $\Pi_n = \{p \mid p$ is a prime > 2, p does not divide n, and $n \equiv c_p^2 \pmod{p}$ for some $c_p \in Z\}$. Let G be the subgroup of $Qx_1 \oplus Qx_2$ generated by $\{x_1, x_2, (x_1 + c_p x_2)/p \mid p \in \Pi_n\}$.

(a) Show that G is homogeneous of type equal to type(Z).

(b) Show that RT(G) is the equivalence class of $\oplus\{Z/pZ \mid p \in \Pi_n\}$ and that OT(A) = $[(m_p)]$, where $m_p = 1$ if $p \in \Pi_n$ and $m_p = 0$ otherwise.

(c) Show that $x_1 \to nx_2$ and $x_2 \to x_1$ induces $f \in E(G)$ with $f^2 = n$.

(d) Show that $Z[\sqrt{n}] = \{a + b\sqrt{n} \quad a, b \in Z\}$ is a subring of $E(G)$.

(e) Show that $E(G) \subseteq Q(\sqrt{n})$. (Hint: Use Lemma 1.12, (a), and (b) to prove that every $0 \neq f \in E(G)$ is a monomorphism and then show that the additive group of $E(G)$ is isomorphic to a subgroup of G and rank(E(G)) = 2.)

(f) Finally, prove that $E(G) = Z[\sqrt{n}]$ and that G is strongly indecomposable.

2.4 Verify the details omitted from the proofs of Example 2.9, Example 2.10, and Example 2.11.

2.5 (R.S. Pierce) For each $0 \le n \in Z$ let $F_n = Z^2$ and $f_n : F_n \to F_{n+1}$ be given by the matrix $\begin{pmatrix} 1 & \epsilon_n \\ 0 & 2 \end{pmatrix}$ where $\epsilon_0 = 1$ and $\epsilon_n = 0$ or $\epsilon_n = 1$ if $n \ge 1$. Define A to be the direct limit of $\{F_n, f_n\}$.

(a) Prove that A is isomorphic to the group constructed in Example 2.7 for the case that $r = 2$ and $\sigma_2 = \Sigma \epsilon_n 2^n \in Z_2^*$.

(b) Deduce that if α_2 is transcendental over Q then A is strongly indecomposable and $E(A) \simeq Z$.

(c) Prove that $Z_2^* \otimes_Z A \simeq Z_2^* \oplus Z_2^*\alpha_2$.

(d) Use (c) and the fact that Z_2^* contains an uncountable set of elements algebraically independent over Q to prove that the set of isomorphism classes of groups constructed as in (a) is uncountable.

2.6 Look up the proof of Theorem 2.13 to verify that p-rank(A) = p-rank(R) for each prime p of Z.

§3. Endomorphism Rings and Decompositions of Rank 2 Groups

The class of rank 2 torsion free groups is a special class of groups, since the rank is small enough so that typeset(A) provides information about decompositions of A and the structure of E(A). Note that if rank(A) = 2 then A is either almost completely decomposable or strongly indecomposable.

For a finite rank torsion free group A, define QE(A) = $Q \otimes_Z E(A)$, a finite dimensional Q-algebra, since E(A) has finite rank. Regard E(A) as a subring of QE(A), via $f \to 1 \otimes f$.

Theorem 3.1: If A is a rank 2 almost completely decomposable group then A is in precisely one of the following classes.

(a) A is homogeneous, $A = B \oplus C$ with type(B) = type(C), $E(A) \simeq \mathrm{Mat}_2(E(B))$ and $QE(A) \simeq \mathrm{Mat}_2(Q)$;

(b) $A = B \oplus C$ with type(B) < type(C), $E(A) \simeq \left\{\begin{pmatrix} f & o \\ h & g\end{pmatrix} \middle| f \in E(B),\ g \in E(C),\right.$ and $\left. h \in \mathrm{Hom}(B,C)\right\}$; and $QE(A) = \left\{\begin{pmatrix} x & o \\ y & z\end{pmatrix} \middle| x, y, z \in A\right\}$.

(c) $A = B \oplus C$ with type(B) and type(C) incomparable, $E(A) \simeq E(B) \times E(C)$, and $QE(A) \simeq QxQ$;

(d) A is indecomposable; there is $0 \neq k \in Z$ and pure rank 1 subgroups B and C of A with incomparable types such that $kA \subseteq B \oplus C \subseteq A$; $A/(B \oplus C)$ is a finite cyclic group; there is a ring monomorphism $\phi : E(A) \to E(B) \times E(C)$ given by $\phi(f) = (f|_B, f|_C)$ such that $k(E(B) \times E(C)) \subseteq \mathrm{Image}(\phi) \subseteq E(B) \times E(C)$; and $QE(A) \simeq QxQ$.

Proof: Exercise 3.3. ///

Let A be a finite rank torsion free group, τ a type, and define $A(\tau) = \{a \in A \mid \mathrm{type}_A(a) \geq \tau\}$. Then $A(\tau)$ is a pure fully invariant subgroup of A. If $\tau, \sigma \in$ typeset (A) then $A(\tau) \subseteq A(\sigma)$ iff $\tau \geq \sigma$. Furthermore, $A(\tau) = A(\sigma)$ iff $\tau = \sigma$.

If rank A = n and if $\tau_1, \ldots, \tau_m \in$ typeset(A) with $\tau_1 < \tau_2 < \ldots < \tau_m$ then $m \leq n$ since $0 \neq A(\tau_m) \subset A(\tau_{m-1}) \subset \ldots \subset A(\tau_1) \subseteq A$. Moreover, m = n implies that $\tau_1 = IT(A)$ since, in this case, $A(\tau_1)$ is a pure subgroup of A with $\mathrm{rank}(A(\tau_1)) = n$.

Theorem 3.2 (Beaumont-Pierce [2]): Assume that A is a torsion free group of rank 2.

(a) If $|\text{typeset}(A)| = 1$ then A is either homogeneous strongly indecomposable or else $A = B \oplus C$ with $\text{type}(B) = \text{type}(C)$.

(b) If $|\text{typeset}(A)| = 2$ then either A is strongly indecomposable or else $A = B \oplus C$ with $\text{type}(B) < \text{type}(C)$.

(c) If $|\text{typeset}(A)| = 3$ then either A is strongly indecomposable or there is $0 \neq k \in Z$ and pure rank 1 subgroups B and C of A with $kA \subseteq B \oplus C \subseteq A$ such that B and C have incomparable types.

(d) If $|\text{typeset}(A)| > 3$ then A is strongly indecomposable. In fact, if X is a rank 1 group with $\text{type}(X) = IT(A)$ then $E(A) \simeq E(X)$ and $QE(A) \simeq Q$.

Proof. (a), (b) are consequences of Theorem 3.1 and the observation that a rank 2 group is either almost completely decomposable or strongly indecomposable.

(c) Let $\text{typeset}(A) = \{\tau_0, \tau_1, \tau_2\}$ where $\tau_0 = IT(A) \in \text{typeset}(A)$ (Proposition 1.7). Then τ_1 and τ_2 are incomparable since otherwise, for example, there is a chain $\tau_0 < \tau_1 < \tau_2$ of types of length 3 in the typeset of a rank 2 group.

Let $B = A(\tau_1)$, $C = A(\tau_2)$ so that $B \oplus C \subseteq A$ noting that $B \cap C = 0$. If A is not strongly indecomposable then $A/(B\oplus C)$ must be finite.

(d) Since $\text{rank}(A) = 2$ and $|\text{typeset}(A)| \geq 4$ there must be at least three distinct types τ_1, τ_2, and τ_3 in $\text{typeset}(A)$ with $\text{rank}(A(\tau_i)) = 1$ for $1 \leq i \leq 3$ (if $\tau \in \text{typeset}(A)$ and $\tau \neq IT(A)$ then $\text{rank}(A(\tau)) = 1$). Now $A/(A(\tau_1) \oplus A(\tau_2))$ is torsion and each $A(\tau_i)$ is pure and fully invariant in A. As in the proof of Example 2.8, $E(A) \subseteq E(A(\tau_1)) \cap E(A(\tau_2)) \subseteq Q$.

In particular, A is strongly indecomposable. If X is a rank-1 group with $\mathrm{type}(X) = \mathrm{IT}(A)$ then $E(A(\tau_1)) \cap E(A(\tau_2)) = E(X)$, since $\mathrm{type}(X) = \inf\{\tau_1,\tau_2\}$. Finally, $E(X) \subseteq E(A)$ since $\mathrm{type}(X) = \mathrm{IT}(A) \leq \mathrm{type}_A(a)$ for each $0 \neq a \in A$. ///

Theorem 3.3: Suppose that A is a strongly indecomposable rank-2 torsion free group. Then A is in precisely one of the following classes:

(a) $|\mathrm{typeset}(A)| = 1$ and $QE(A)$ is a quadratic number field (a field with Q-dimension ≤ 2);

(b) $|\mathrm{typeset}(A)| = 2$ and $QE(A) \simeq Q$;

(c) $|\mathrm{typeset}(A)| = 2$ and $QE(A) \simeq \left\{\begin{pmatrix} x & 0 \\ y & x \end{pmatrix} \mid x, y \in Q\right\}$

(d) $|\mathrm{typeset}(A)| \geq 3$ and $QE(A) \simeq Q$.

In each case, $E(A)$ is commutative.

Proof.

Since A is strongly indecomposable, rank $E(A) < 4 = 2^2$ (Corollary 1.13). Thus $QE(A)$ is an artinian Q-algebra with $\dim_Q QE(A) < 4$. Let J = Jacobson radical of $QE(A)$. Since $QE(A)$ is artinian, J is nilpotent. Moreover, $QE(A)/J$ is a semi-simple artinian algebra with $\dim_Q J < \dim_Q QE(A)$. In fact, since A is strongly indecomposable, $QE(A)/J$ is a division algebra (Corollary 7.8). It is proved in Lemma 10.7 that Q-division algebras have square dimension over their center, a field. Consequently, $QE(A)/J$ must be a quadratic number field.

(a) Suppose that $|\mathrm{typeset}(A)| = 1$, i.e. A is homogeneous. Then it is sufficient to prove that if $0 \neq f \in E(A)$ then f is a monomorphism; in which case $J = 0$ and $QE(A)$ is a quadratic number field. Assume that $0 \neq f \in E(A)$ and $0 \neq a \in A$ with $f(a) = 0$. Then $f : A/\langle a\rangle_* \to A$. If $b \in A\backslash\langle a\rangle_*$ then $\mathrm{type}(A) = \mathrm{type}_A(b) \leq \mathrm{type}_A(f(b)) = \mathrm{type}(A)$. As a consequence of Baer's Lemma (Lemma 1.12), $\langle a\rangle_*$ is a summand of A contradicting the assumption that A is strongly indecomposable. This proves (a).

(b) (c) Suppose that $|\text{typeset}(A)| = 2$, say B and C are pure rank 1 subgroups of A with $\text{type}(B) = IT(A) < \text{type}(C)$ and $\text{typeset}(A) = \{\text{type}(B), \text{type}(C)\}$. Now $A/(B \oplus C)$ is torsion so $QE(A) \subseteq \left\{ \begin{pmatrix} f & 0 \\ g & h \end{pmatrix} \middle| f : QB \to QB, h : QC \to QC, g : QB \to QC \right\}$, noting that $\text{Hom}(C,B) = 0$. Thus, $QE(A) \subseteq \left\{ \begin{pmatrix} x & 0 \\ y & z \end{pmatrix} \middle| x, y, z \in Q \right\}$. Since $QE(A)/J$ is a quadratic number field it follows that $QE(A)/J \simeq Q$, i.e., $QE(A) \subseteq \left\{ \begin{pmatrix} x & 0 \\ y & x \end{pmatrix} \middle| x, y \in Q \right\}$ and $J \subseteq \left\{ \begin{pmatrix} 0 & 0 \\ y & 0 \end{pmatrix} \middle| y \in Q \right\}$. If $J = 0$ then $QE(A) \simeq Q$. Otherwise, $QE(A) = \left\{ \begin{pmatrix} x & 0 \\ y & x \end{pmatrix} \middle| x, y \in Q \right\}$.

(d) is Theorem 3.2.d in case $|\text{typeset}(A)| > 3$. Assume that $\text{typeset}(A) = \{\tau_0, \tau_1, \tau_2\}$ with $\tau_0 = IT(A) \in \text{typeset}(A)$. Then $B = A(\tau_1)$ and $C = A(\tau_2)$ are pure rank-1 fully invariant subgroups of A. Now $\phi : E(A) \to E(B) \times E(C) \subseteq Q \times Q$, defined by $\phi(f) = f|_B \oplus f|_C$ is a monomorphism. Since A is strongly indecomposable, it follows that $QE(A) \simeq Q$. ///

EXERCISES

3.1 (J. Reid): This exercise generalizes Theorem 3.2.d. Let A be a torsion free group of rank n having 2n-1 distinct types $\{\tau_1, \ldots, \tau_n, \sigma_1, \ldots, \sigma_{n-1}\} \subseteq \text{typeset}(A)$ such that (i) for each $1 \le i \le n$, $\tau_i \not\ge \inf \{\tau_j | j \ne i\}$; (ii) for each $1 \le i \le n-1$, $\sigma_i \not\ge \inf \{\sigma_j | j \ne i\}$; and (iii) if $\sigma_0 = \inf \{\sigma_i | 1 \le i \le n-1\}$ then $\tau \not\ge \sup \{\tau_j, \sigma_0\}$ for each $1 \le j \le n$, $\tau \in \text{typeset}(A)$.

(a) Show that there is $Za_1 \oplus \ldots \oplus Za_n \subseteq A$ with $\text{type}_A(a_i) = \tau_i$ and $A(\tau_i) = \langle a_i \rangle_*$.

(b) Show that $A(\sigma_0)$ is a pure fully invariant subgroup of A with $\text{rank}(A(\sigma_0)) = n-1$ and that $\text{Hom}(A, A(\sigma_0)) = 0$.

(c) Show that E(A) is isomorphic to a subring of Q and deduce that A is strongly indecomposable.

3.2 Show that each of the classes in Theorem 3.1, Theorem 3.2, and Theorem 3.3 is not empty.

3.3 Prove Theorem 3.1 using results from Section 1 and Section 2.

§4. Pure Subgroups of Completely Decomposable Groups:

Let $A = A_1 \oplus \ldots \oplus A_n$ be a completely decomposable group, where each A_i has rank 1. Then A is homogeneous iff there is a type τ with $\text{type}(A_i) = \tau$ for each i (Exercise 1.7). If B is a pure subgroup of a homogeneous completely decomposable group A then B is a summand of A and B is homogeneous completely decomposable with $\text{type}(B) = \text{type}(A)$. (Exercise 1.2).

In general, pure subgroups of completely decomposable groups need not be completely decomposable.

Example 4.1: There is a torsion free group $A = A_1 \oplus A_2 \oplus A_3$ with $\text{rank}(A_i) = 1$ for each i and a pure rank 2 subgroup B of A such that B is strongly indecomposable.

Proof. Let A_1, A_2, A_3 be torsion free rank 1 groups with

$$\text{type}(A_1) = [(\infty, \infty, 0, \infty, \infty, \infty, \ldots)]$$

$$\text{type}(A_2) = [(\infty, 0, \infty, \infty, \infty, \infty, \ldots)]$$

$$\text{type}(A_3) = [(0, \infty, \infty, \infty, \infty, \infty, \ldots)]$$

Let $A = A_1 \oplus A_2 \oplus A_3$, $0 \neq a_i \in A_i$, $b_1 = a_1 + a_2$, $b_2 = a_2 + a_3$ and $B = \langle b_1, b_2 \rangle_*$, a rank-2 group.

Then $$\text{type}_B(b_1) = \inf\{\text{type}_A(a_1), \text{type}_A(a_2)\} = [\infty, 0, 0, \infty, \infty, \infty, \ldots]$$

$$\text{type}_B(b_2) = \inf\{\text{type}_A(a_2), \text{type}_A(a_3)\} = [0, 0, \infty, \infty, \infty, \infty, \ldots]$$

and

$$\text{type}_B(b_1 - b_2) = \inf\{\text{type}_A(a_1), \text{type}_A(a_3)\} = [0, \infty, 0, \infty, \infty, \infty, \ldots]$$

are three distinct incomparable elements of $\text{typeset}(B)$. As a consequence of Theorem 3.2.d, B is strongly indecomposable.

Following is a characterization of the class of pure subgroups of completely decomposable groups given by Butler [1] using techniques originated by Baer [2].

A class C of finite rank torsion-free groups is called a <u>torsion free class</u> if C is closed with respect to finite direct sums, pure subgroups, and torsion free homomorphic images. For example, if τ is a type and C_τ is the class of finite rank homogeneous completely decomposable groups of type τ then C_τ is a torsion free class (Exercise 4.1). If C' is any class of finite rank torsion free groups then there is a smallest torsion free class containing C'.

Let A be a finite rank torsion free group and define $C(A)$ to be the smallest torsion free class containing A. If $\mathrm{rank}(A) = 1$ with $\mathrm{type}(A) = \tau$ then $C(A) = C_\tau$.

<u>Proposition 4.2</u>: Suppose that A is a torsion free group with $\mathrm{rank}(A) > 1$ and that $A = A_1 + \ldots + A_n$, where each A_i is a pure rank-1 subgroup of A. Then $\phi : A \to A/A_1 \oplus \ldots \oplus A/A_n$, defined by $\phi(a) = (a + A_1, \ldots, a + A_n)$ is a monomorphism with pure image.

<u>Proof.</u> Let $K = \mathrm{Kernel}(\phi) = \cap\{A_i \mid 1 \leq i \leq n\}$. If $K \neq 0$ then $K = A_i$ since $\mathrm{rank}(A_i) = 1$, for each i, and $A = K$, contradicting the assumption that $\mathrm{rank}(A) > 1$.

Assume that $\phi(A)$ is not pure in $B = A/A_1 \oplus \ldots \oplus A/A_n$. Then there is a prime p and $a \in A \backslash pA$ with $f(a) = (a + A_1, \ldots, a + A_n) = p(x_1 + A_1, \ldots, x_n + A_n)$ for some $x_i \in A$. Thus $a_i = a - px_i \in A_i \backslash pA_i$ for each i. But $\mathrm{rank}(A_i) = 1$, so $A_i = Za_i + pA_i$ and $A/pA \simeq Z/pZ$ since $A = A_1 + \ldots + A_n = Za_1 + \ldots + Za_n + pA = Za + pA$. Moreover, $(A_i)_p = Z_p a_i \simeq Z_p$ so that A_p is a free Z_p-module, being a finitely generated torsion free Z_p-module. Now $m = Z_p$ - rank A_p = rank $A > 1$ and $Z/pZ \simeq A/pA \simeq A_p/pA_p \simeq (Z/pZ)^m$, a contradiction.

Proposition 4.3: Assume that A is torsion free of finite rank > 1 and that $A = A_1 + A_2 + \ldots + A_n$, where each A_i is a pure rank-1 subgroup of A. For a subset I of $n^+ = \{1, 2, \ldots, n\}$, let $A(I) = \Sigma\,\{A_i \mid i \in I\}$. If B is a pure subgroup of A with $\text{rank}(B) \geq 2$ then $B = \Sigma\,\{B \cap A(I) \mid I \subsetneq n^+\}$.

Proof. Let $A' = A_1 \oplus \ldots \oplus A_n \xrightarrow{f} A \to 0$ be an epimorphism. Then $B' = f^{-1}(B)$ is a pure subgroup of A'. If $B' = \Sigma\,\{B' \cap A'(I) \mid I \subsetneq n^+\}$ then $B = f(B') = \Sigma\,\{B \cap A(I) \mid I \subsetneq n^+\}$. Thus, it is sufficient to assume that $A = A_1 \oplus \ldots \oplus A_n$.

If p is a prime then B_p is a pure subgroup of A_p. Therefore, it is sufficient to assume that A is p-local since if $B_p = \Sigma\,\{B_p \cap A_p(I) \mid I \subsetneq n^+\}$ then $B = \cap_p B_p = \Sigma\,\{B \cap A(I) \mid I \subsetneq n^+\}$.

Since $A = A_1 \oplus \ldots \oplus A_n$ is assumed to be p-local, each A_i is isomorphic to Z_p or Q (consider the possibilities for type(A_i)). Write $A = F \oplus D$, where F is a free Z_p-module and D is a Q-vector space. Then $B = B_1 \oplus B_2$, where $B_2 = D \cap B$ is a free Q-module and B_1 is a free Z_p-module (B/B_2 is isomorphic to a submodule of the free Z_p-module $F \simeq A/D$).

Write $A = F \oplus D = (Z_p a_1 \oplus \ldots \oplus Z_p a_m) \oplus (Qa_{m+1} \oplus \ldots \oplus Qa_n)$, $B = B_1 \oplus B_2 = (Z_p b_1 \oplus \ldots \oplus Z_p b_k) \oplus (Qb_{k+1} \oplus \ldots \oplus Qb_\ell)$, and $b_i = r_{i,1}a_1 + \ldots + r_{i,m}a_m + q_{i,m+1}a_{m+1} + \ldots + q_{i,n}a_n$ with $r_{i,j} \in Z_p$ and $q_{i,j} \in Q$. If $1 \leq i \leq k$ then $h_p(b_i) = 0$ so some $r_{ij} \in Z_p \backslash pZ_p$ is a unit of Z_p. Moreover, if $k + 1 \leq i \leq \ell$ then $r_{i,j} = 0$ for $1 \leq j \leq m$, since $h_p(b_i) = \infty$, and $b_i \in B_2 \subseteq D$; and some $q_{ij} \neq 0$ is a unit of Q. Perform row operations on the matrix (r_{ij}, q_{ij}) so that $B = Z_p b_1' \oplus \ldots \oplus Z_p b_k' \oplus Qb_{k+1}' \oplus \ldots \oplus Qb_\ell'$ with each $b_i' \in B \cap A(I)$ for some $I \subsetneq n^+$, noting that $\ell = \text{rank}(B) > 1$. ///.

Remark: Let $B = Z(x+y) \subseteq A = Zx \oplus Zy = A_1 \oplus A_2$ to see that the hypothesis $\text{rank}(B) \geq 2$ in Proposition 4.3 is necessary.

Theorem 4.4: Let $\underline{P}$ be the class of pure subgroups of finite rank completely decomposable groups, $\underline{H}$ the class of torsion free homomorphic images of finite rank completely decomposable groups, and let R be the smallest torsion free class containing the class of all rank-1 torsion free groups. Then $P = H = R$.

Proof. Clearly $H \subseteq R$ and $P \subseteq R$. To show that $H = R$ it is sufficient to prove that if A is in H and if B is a pure subgroup of A with rank(B) ≥ 2 then B is in H. Note that if A is in H then there is a least integer $n(A) \geq 1$ with $A = A_1 + \ldots + A_{n(A)}$ and each A_i a pure rank-1 subgroup of A (if A_i is a rank-1 subgroup of A then $<A_i>_*$ is a pure rank-1 subgroup of A). For each $I \subsetneq n(A)^+$, $B(I) = B \cap A(I)$ is a pure subgroup of $A(I)$ and $A(I)$ is an H-group with $n(A(I)) < n(A)$. By induction on $n(A)$, $B(I)$ is in H. Moreover, $B = \Sigma \{B(I) | I \subsetneq n(A)^+\}$ (Proposition 4.3) so that B is in H proving that $H = R$.

To show that $R = P$ it is sufficient to prove that if A is an R-group then A is isomorphic to a pure subgroup of a completely decomposable group. Let A be an R-group. Then $A = A_1 + \ldots + A_n$, where each A_i is a pure rank-1 subgroup of A, since $R = H$. For each i, $B_i = A/A_i$ is an R-group with rank$(B_i) <$ rank(A). By induction on rank(A), each B_i is a pure subgroup of a completely decomposable group. But A is isomorphic to a pure subgroup of $B = B_1 \oplus \ldots \oplus B_n$ (Proposition 4.2) which, in turn, is isomorphic to a pure subgroup of a completely decomposable group. Thus, $R = P$. ///

Corollary 4.5: Homogeneous R-groups are completely decomposable.

Proof. If A is in $R = H$ then $A = A_1 + \ldots + A_n$ where each A_i is a pure rank-1 subgroup of A. But A is homogeneous of type τ so that $A \in C_\tau$ whence A is homogeneous completely decomposable. ///

Let A be a finite rank torsion free group and let τ be a type. Recall that $A(\tau) = \{a \in A | \text{type}_A(a) \geq \tau\}$. Define $A^*(\tau)$ to be the subgroup of A generated by $\{a \in A | \text{type}_A(a) > \tau\}$. Then $A(\tau)$ and $A^*(\tau)$ are fully invariant subgroups of A. Note that $<A^*(\tau)>_* \subseteq A(\tau)$, since $A(\tau)$ is pure in A.

Theorem 4.6: Let A be a finite rank torsion free group. Then A is an R-group iff (i) typeset(A) is finite; (ii) for each type τ, $A(\tau)/\langle A^*(\tau)\rangle_*$ is homogeneous completely decomposable of type τ; and (iii) for each type τ, $\langle A^*(\tau)\rangle_*/A^*(\tau)$ is finite.

Proof. ($\rightarrow$) (i) Since $R = P$, A is a pure subgroup of a completely decomposable group D. Thus, typeset(A) $\subseteq$ typeset(D), which is finite.

(ii) Since A is an R-group, $A(\tau)$ and $A(\tau)/\langle A^*(\tau)\rangle_* = B$ are R-groups. In view of Corollary 4.5, it is sufficient to prove that B is homogeneous of type τ. Let $f : A(\tau) \rightarrow B \rightarrow 0$ be the quotient map and let Y be a pure rank-1 subgroup of B. Then $f^{-1}(Y)$ is a pure subgroup of $A(\tau)$, hence an R-group. Thus, $f^{-1}(Y) = A_1 + \ldots + A_n$ where each A_i is a pure rank-1 subgroup of A with $\text{type}(A_i) \geq \tau$. If type $(A_i) > \tau$ then $A_i \subseteq \langle A^*(\tau)\rangle_*$ so that $f(A_i) = 0$. Therefore, $Y = f(A_1) + \ldots + f(A_m)$ with $\text{type}(A_i) = \tau$ for each i. But $\text{type}(Y) = \sup\{\text{type } f(A_i)\} = \tau$, (Exercise 1.6) as desired.

(iii) Let $B = \langle A^*(\tau)\rangle_*$ so that $B^*(\tau) = A^*(\tau)$ and assume that $B/B^*(\tau) \neq 0$. Then B contains an element of type τ, else $B = B^*(\tau)$, and B is an R-group with $B/B^*(\tau)$ torsion. Write $B = B_1 + \ldots + B_k + B^*(\tau)$, where each B_i is a pure rank-1 subgroup of B with $\text{type}(B_i) = \tau$. Now $C_i = B_i \cap B^*(\tau)$ is a pure rank-1 subgroup of $B^*(\tau)$ so that $\text{type}(C_i) = \tau$ since $C_i \subseteq B_i$, $\text{type}(B_i) = \tau$, and $\text{type}(C_i) \geq \tau$. Consequently, there is $0 \neq n \in Z$ with $nB_i \subseteq C_i \subseteq B_i$ for each i. Thus, $nB = nB_1 + \ldots + nB_k + nB^*(\tau) \subseteq C_1 + \ldots + C_k + B^*(\tau) = B^*(\tau)$ so that $B/B^*(\tau) = \langle A^*(\tau)\rangle_*/A^*(\tau)$ is finite.

($\leftarrow$) For each type τ, $A(\tau) = \langle A^*(\tau)\rangle_* \oplus C_\tau$, where $C_\tau \simeq A(\tau)/\langle A^*(\tau)\rangle_*$ is homogeneous completely decomposable of type τ (a consequence of Baer's Lemma, Lemma 1.12). Since typeset(A) is finite, $C_\tau = 0$ for all but a finite number of types $\tau_1, \ldots, \tau_n$ in typeset(A). Let $C = C_{\tau_1} + \ldots + C_{\tau_n}$. Each C_{τ_i}, hence C, is an R-group.

It suffices to prove that A/C is finite; in which case there is an epimorphism $Z^k \oplus C \rightarrow A \rightarrow 0$ for some $0 \neq k \in Z$ and A is an R-group.

For each $\tau \in$ typeset(A) define $d(\tau)$ to be the length of a maximal chain of types $\tau = \tau_0 < \tau_1 < \ldots < \tau_{d(\tau)}$ in typeset(A). Define $A^\tau = \Sigma\{C_\sigma | \sigma \geq \tau\}$. Note that $A^\tau = C$ if $\tau = IT(A)$. If $d(\tau) = 0$ then τ is a maximal type in typeset(A), hence $A^*(\tau) = 0 = \langle A^*(\tau)\rangle_*$ and $A^\tau = C_\tau = A(\tau)$. Next assume that $d(\tau) = d \geq 1$ and that $A(\sigma)/A^\sigma$ is finite for all $\sigma \in$ typeset(A) with $d(\sigma) < d$. Now $A^*(\tau) = \Sigma\{A(\sigma) | \sigma > \tau\}$ and $d(\sigma) < d$ if $\sigma > \tau$ so that $A^\tau = \Sigma\{C_\sigma | \sigma \geq \tau\} = \Sigma\{C_\sigma | \sigma > \tau\} \oplus C_\tau$ has finite index in $A^*(\tau) \oplus C_\tau$ by induction on $d(\tau)$. Since $\langle A^*(\tau)\rangle_*/A^*(\tau) \simeq A(\tau)/(A^*(\tau) \oplus C_\tau)$ is finite, $A(\tau)/A^\tau$ is finite for all $\tau \in$ typeset(A). In particular, if $\tau = IT(A) \in$ typeset(A) then $A/C = A(\tau)/A^\tau$ is finite. ///

Included in the proof of Theorem 4.6 is the proof of the following:

Corollary 4.7. If A is an R-group and if B is a subgroup of finite index in A then B is an R-group.

While not every finite rank torsion free group is an R-group the following theorem, due to Brenner-Butler [1], shows that the class of R-groups is relatively large in the class of finite rank torsion free groups.

Theorem 4.8. If K is a Q-algebra of dimension n over Q then there is an R-group A with rank(A) = 2n and QE(A) $\simeq$ K.

EXERCISES

4.1 Prove that if τ is a type and C_τ is the class of finite rank homogeneous completely decomposable groups of type τ then C_τ is a torsion free class. Moreover, if X is a rank-1 torsion free group of type τ prove that $C(X) = C_\tau$.

4.2 Prove that if A is an R-group and if typeset(A) is totally ordered then A is completely decomposable. Deduce a special case of Theorem 2.3: if $kA \subseteq A_1 \oplus \ldots \oplus A_n \subseteq A$ for some $0 \neq k$ where each A_i has rank-1 and $\text{type}(A_1) \leq \text{type}(A_2) \leq \ldots \leq \text{type}(A_n)$ then A is completely decomposable.

4.3 Prove that if $A = A_1 \oplus \ldots \oplus A_n$ is a torsion free group of finite rank and τ is a type then $A(\tau) = A_1(\tau) \oplus \ldots \oplus A_n(\tau)$ and $A^*(\tau) = A_1^*(\tau) \oplus \ldots \oplus A_n^*(\tau)$.

4.4 For a torsion free abelian group A and a type τ define $w_A(\tau) = \dim_Q Q\otimes_Z (A(\tau)/A^*(\tau))$. Prove that if A and B are (finite rank) completely decomposable groups then $A \simeq B$ iff $w_A(\tau) = w_B(\tau)$ for each type τ.

4.5 Let A be a finite rank torsion free group. Prove that A is an R-group iff (i) typeset(A) is finite and (ii) for each type τ there is a homogeneous completely decomposable group C of type τ such that $A(\tau)$ is quasi-isomorphic to $A^*(\tau) \oplus C$.

4.6 Show that the groups constructed in Examples 2.2, 2.4, 2.9, 2.10, and 2.11 are R-groups.

§5. Homogeneous Completely Decomposable Groups and Generalizations:

Let A be a finite rank torsion free group and define $P(A)$ to be the category of summands of finite direct sums of copies of A and $P^{\infty}(A)$ to be the category of summands of arbitrary direct sums of copies of A. For example, if rank(A) = 1 with type(A) = τ then $P(A)$ is the category of finite rank homogeneous completely decomposable groups of type τ.

For a ring R, let $P(R)$ be the category of finitely generated projective right R-modules and $P^{\infty}(R)$ be the category of projective right R-modules.

Suppose that A is an abelian group and regard A as a left E(A)-module. If G is another abelian group then $H_A(G) = \mathrm{Hom}(A,G)$ has the structure of a right E(A)-module, where if $r \in E(A)$, $f \in \mathrm{Hom}(A,G)$ then $fr \in \mathrm{Hom}(A,G)$ is defined by $(fr)(a) = (f \circ r)(a)$. Furthermore, if $g \in \mathrm{Hom}(G, G')$ then $H_A(g) : H_A(G) \to H_A(G')$ defined by $H_A(g)(f) = gf$ is a right E(A)-homomorphism. Therefore, H_A is a functor from $P(A)$ to $M_{E(A)}$, the category of right E(A)-modules, since $H_A(1_G) = 1_{H_{A}(G)}$ and $H_A(gg') = H_A(g)H_A(g')$.

If $M \in M_{E(A)}$ define $T_A(M) = M \otimes_{E(A)} A$, an abelian group. Then T_A is a functor from $P(E(A))$ to the category, Ab, of abelian groups, where if $f : M \to M'$ then $T_A(f) = f \otimes 1 : T_A(M) \to T_A(M')$.

Theorem 5.1: (a) If A is an abelian group then $H_A : P(A) \to P(E(A))$ is a category equivalence with inverse $T_A : P(E(A)) \to P(A)$.

(b) If A is finite rank torsion free then $H_A : P^{\infty}(A) \to P^{\infty}(E(A))$ is a category equivalence with inverse $T_A : P^{\infty}(E(A)) \to P^{\infty}(A)$.

Proof. If $B \oplus C \simeq A^n$ then $H_A(B) \oplus H_A(C) \simeq H_A(A^n)$, a free E(A)-module so that $H_A(B) \in P(E(A))$. Similarly, $T_A: P(E(A)) \to P(A)$ and $T_A : P^{\infty}(E(A)) \to P^{\infty}(A)$, since $(\oplus M_i) \otimes_{E(A)} A \simeq \oplus (M_i \otimes_{E(A)} A)$. Suppose that A is finite rank torsion free with maximal independent subset $x_1, \ldots, x_n$ and that $B \oplus C \simeq \oplus_I A$ for some index set I. Then $H_A(B) \oplus H_A(C) \simeq H_A(\oplus_I A)$.

But $H_A(\oplus_I A) \simeq \oplus_I H_A(A)$, in this case, noting that if $f : A \to \oplus_I A$ there is a finite subset J of I with $f(x_i) \in \oplus_J A$ for $1 \le i \le n$ so that $f(A) \subseteq \oplus_J A$. Thus, $H_A : P^\infty(A) \to P^\infty(E(A))$.

(a) To prove that H_A is a category equivalence with inverse T_A it is sufficient to prove that for each $B \in P(A)$ there is a natural group isomorphism $\theta_B : T_A H_A(B) \to B$ and for each $M \in P(E(A))$ there is a natural $E(A)$-isomorphism $\phi_M : M \to H_A T_A(M)$.

For $B \in P(A)$ define $\theta_B : T_A H_A(B) \to B$ by $\theta_B(f \otimes a) = f(a)$, a well defined natural homomorphism. Note that $\theta_A : \mathrm{Hom}(A,A) \otimes_{E(A)} A \to A$ is an isomorphism from which it follows that $\theta_G : \mathrm{Hom}\,(A,G) \otimes_{E(A)} A \to G$ is an isomorphism whenever $G = A^n$. Now assume that $G = A^n$ and that $\delta : G \to B \oplus C$ is an isomorphism. Then there is a commutative diagram

$$\begin{array}{ccc} T_A H_A(G) & \xrightarrow{\ \sigma\ } & T_A H_A(B) \oplus T_A H_A(C) \\ \Big\downarrow \theta_G & & \Big\downarrow \theta_B \oplus \theta_C \\ G & \xrightarrow{\ \delta\ } & B \oplus C \end{array}$$

where $\sigma(f \otimes a) = (\pi_B \delta f \otimes a, \pi_C \delta f \otimes a)$, and $\pi_B : B \oplus C \to B$, $\pi_C : B \oplus C \to C$ are projections with $1 = \pi_B + \pi_C$. Furthermore, σ is an isomorphism with inverse $\sigma' : T_A H_A(B) \oplus T_A H_A(C) \to T_A H_A(G)$ given by $\sigma'(f_1 \otimes a_1, f_2 \otimes a_2) = \delta^{-1} f_1 \otimes a_1 + \delta^{-1} f_2 \otimes a_2$. Therefore, θ_B must be an isomorphism since θ_G is an isomorphism.

For $M \in P(E(A))$, define $\phi_M : M \to H_A T_A(M)$ by $\phi_M(m)(a) = m \otimes a$. Then ϕ_M is a well defined natural $E(A)$ - homomorphism. But $\phi_{E(A)}$ is an isomorphism from which it follows that ϕ_M is an isomorphism whenever $M = E(A)^n$ is a free $E(A)$-module. If $M \oplus N \simeq E(A)^n$ then ϕ_M is an isomorphism, by an argument analogous to that of the preceeding paragraph.

(b) The proof of (b) is the same as the proof of (a), the hypothesis that A is finite rank torsion free is used only to show that $H_A : P^\infty(A) \to P^\infty(E(A))$ is well defined. ///

Remark: Theorem 5.1 is proved in an additive category setting in Section 7.

Corollary 5.2: Suppose that A is finite rank torsion free. Every B in $P^{\infty}(A)$ (with rank $B < \infty$) is isomorphic to a direct sum of copies of A iff every (finitely generated) projective right $E(A)$-module is free. ///

If $E(A)$ is a principal ideal domain, e.g. $E(A)$ is isomorphic to a subring of Q, then every (finitely generated) projective $E(A)$-module is free. Thus, every $B \in P(A)$ is isomorphic to a direct sum of copies of A (Corollary 5.2). Examples of groups with $E(A)$ a principal ideal domain are given in Section 2.

Corollary 5.3: If B is a summand of a homogeneous completely decomposable group of type τ then B is homogeneous completely decomposable of type τ. ///

Theorem 5.1 and Corner's Theorem (Theorem 2.13) may be used to construct examples of finite direct sums of strongly indecomposable groups with non-equivalent direct sum decompositions, in contrast to Examples 2.9-2.12 which are direct sums of indecomposable almost completely decomposable groups.

Example 5.4: There are strongly indecomposable torsion free groups A, B, C each with rank 4 such that $B \oplus C \simeq A \oplus A$ but $B \not\simeq A$ and $C \not\simeq A$.

Proof. Let $R = Z[\sqrt{-5}] = \{a + b\sqrt{-5} \mid a, b \in Z\}$ and let $I_1 = 3R + (1+\sqrt{-5})R$, $I_2 = 3R + (1-\sqrt{-5})R$, two proper ideals of R with $I_1 \cap I_2 = 3R \simeq R$. Then I_1 and I_2 are not principal ideals, a consequence of the fact that $N(rs) = N(r)N(s)$ for each $r, s \in R$ where $N(a + b\sqrt{-5}) = a^2 + 5b^2$ is the norm of $a + b\sqrt{-5}$.

There is an exact sequence of R-modules $0 \to I_1 \cap I_2 \to I_1 \oplus I_2 \to R = I_1 + I_2 \to 0$, noting that $1 \in I_1 + I_2$. Thus $I_1 \oplus I_2 \simeq R \oplus (I_1 \cap I_2) \simeq R \oplus R$. By Corner's Theorem, $R \simeq E(A)$ for some A of rank $4 = 2$ rank (R). Furthermore, $A \oplus A \simeq B \oplus C$, where $B = I_1 \otimes_{E(A)} A$ and $C = I_2 \otimes_{E(A)} A$.

The group A is strongly indecomposable, since $E(A)$ is a domain. Moreover, $B \neq A$ and $C \neq A$ since, for example, $H_A(B) \simeq I_1 \neq R$. Also B and C are strongly indecomposable since, for example, R/I_1 finite implies that $A/B \simeq (E(A) \otimes_{E(A)} A)/(I_1 \otimes_{E(A)} A)$ is finite and A is strongly indecomposable. ///

There is a natural generalization of the notion of $B(\tau)$ for a type τ and a finite rank torsion free group B. Let A be a finite rank torsion free group and G an abelian group. Define the A-socle of G, $S_A(G)$, to be the image of the evaluation map $\mathrm{Hom}(A,G) \otimes A \to G$, i.e., $S_A(G)$ is the subgroup of G generated by $\{f(A) \mid f \in \mathrm{Hom}(A,G)\}$. If $B \in P(A)$ then $S_A(B) = B$ since B is a summand of A^n for some n.

Proposition 5.5: Let G be a finite rank torsion free group and let A be a rank-1 torsion free group with $\mathrm{type}(A) = \tau$. Then $S_A(G) = G(\tau)$. Furthermore, $S_A(G) = G$ iff $\tau \le IT(G)$.

Proof. If $0 \neq a \in A$ and $0 \neq f \in \mathrm{Hom}(A,G)$ then $\tau = \mathrm{type}_A(a) \le \mathrm{type}_G(f(a))$ so that $S_A(G) \subseteq G(\tau)$. Conversely, if $x \in G$ with $\mathrm{type}_G(x) \ge \tau$ then there is $f : A \to \langle x \rangle_* \subseteq G$ induced by $f(a) = x$, where $a \in A$ with $h^A(a) \le h^G(x)$. Thus $G(\tau) \subseteq S_A(G)$. Finally, $S_A(G) = G$ iff $G(\tau) = G$ iff $\tau \le IT(G)$. ///

The hypotheses of Baer's Lemma (Lemma 1.12) are that $A = G/B$ is a rank-1 torsion free group with $\mathrm{type}(A) = \tau$ and that $\mathrm{type}_G(x) = \tau$ for each $x \in G \backslash B$. Therefore, $G(\tau) + B = G = S_A(G) + B$, as a consequence of Proposition 5.5. The following theorem is a generalization of Baer's Lemma.

Theorem 5.6: Suppose that A is finite rank torsion free. Then the following are equivalent: (a) If B is a subgroup of an abelian group G with $G/B \simeq A$ and if $S_A(G) + B = G$ then B is a summand of G; (b) If I is a maximal right ideal of $E(A)$ then $IA \neq A$.

Proof. (b) ⟹ (a) Let $\pi : G \to A \to 0$ be an epimorphism with $\mathrm{Kernel}(\pi) = B$. Define $I = \{\pi h \mid h \in \mathrm{Hom}(A,G)\}$ a right ideal of $E(A)$. Since $S_A(G) + B = G$, $IA = A$. By (b), $I = E(A)$, otherwise I is contained in a maximal right ideal J with $JA = A$. Thus there is $h \in \mathrm{Hom}(A,G)$ with $\pi h = 1_A$ so that $G = h(A) \oplus B$.

(a) ⟹ (b) Let I be a right ideal of $E(A)$ with $IA = A$. Choose a free $E(A)$-module P and an epimorphism $\pi : P \to I \to 0$. As a consequence of Theorem 5.1 there is a commutative diagram

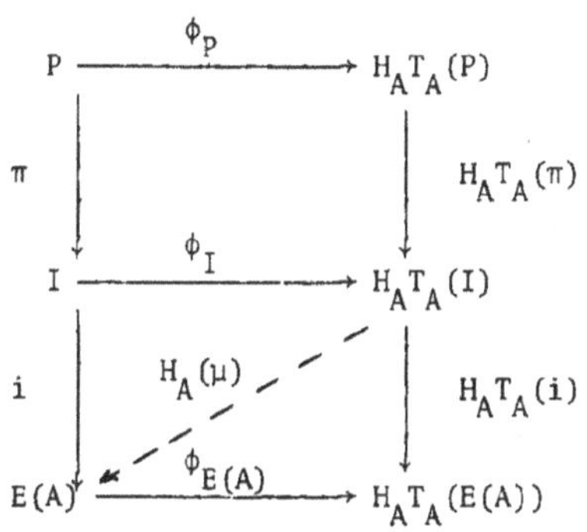

where i is inclusion and ϕ_P, $\phi_{E(A)}$ are isomorphisms.

Let $\mu : I \otimes_{E(A)} A \to IA = A$ be given by $\mu(r \otimes a) = ra$. Then $H_A(\mu) : H_AT_A(I) = \mathrm{Hom}(A, I \otimes_{E(A)} A) \to E(A)$ and $\phi_{E(A)} H_A(\mu) = H_AT_A(i)$. Thus $H_A(\mu)\, H_AT_A(\pi)\, \phi_P = i\pi$. Now $\mu T_A(\pi) : T_A(P) \to A \to 0$ is an epimorphism with $S_A(T_A(P)) = T_A(P)$, hence there is $\sigma : A \to T_A(P)$ with $\mu T_A(\pi)\sigma = 1_A$, by hypothesis. Thus $H_A(\mu)\, H_AT_A(\pi) = H_A(\mu T_A(\pi)) : H_AT_A(P) \to E(A)$ is an epimorphism. But ϕ_P is an epimorphism so that $i\pi$, hence i, is an epimorphism whence $I = E(A)$. ///

Corollary 5.7: Assume that A is a torsion free group of finite rank and that $E(A)$ is right principal (every right ideal of $E(A)$ is principal). If B is a subgroup of an abelian group G with $G/B \simeq A$ and if $S_A(G) + B = G$ then B is a summand of G.

Proof. Let $I = fE(A)$ be a maximal right ideal of $E(A)$ with $IA = A$. Then $f(A) = A$ so that f is an automorphism of A, since rank $A < \infty$. Therefore, f is a unit of $E(A)$ and $I = E(A)$. Now apply Theorem 5.6. ///

A finite rank torsion free group A is faithful if $(E(A)/I) \otimes_{E(A)} A \neq 0$ for each maximal right ideal I of $E(A)$, i.e., $IA \neq A$ for each maximal right ideal I of $E(A)$. If $E(A)$ is right principal then A is faithful (Corollary 5.7 and Theorem 5.6). Other examples are given in the next theorem.

Lemma 5.8 (Kaplansky [2]). Let R be a commutative ring, A a finitely generated R-module, and I an ideal of R. If $IA = A$ then $(1+y)A = 0$ for some $y \in I$.

Proof. Let $a_1, \ldots, a_n$ be a set of R-generators of A. For each i, $a_i = \Sigma_j r_{ij} a_j$, $r_{ij} \in I$. If $M = I_{n\times n} - (r_{ij})$, where $I_{n\times n}$ is the $n\times n$ identity matrix, then $M(a_1, \ldots, a_n)^t = 0$. Now $\det(M) = 1+y$ for some $y \in I$ and $(\det M)a_i = 0$ for each i, since $(\operatorname{adj} M)M = (\det M)I$, which implies that $(1+y)A = 0$. ///

Theorem 5.9: Let A be a torsion free group of finite rank.

(a) If $E(A)$ is commutative then A is faithful.

(b) If $E(A)$ is right hereditary (every right ideal of $E(A)$ is projective) then A is faithful.

Proof. (a) Let I be an ideal of $E(A)$ with $IA = A$. Then $Q \otimes I$ is an ideal of the commutative ring $Q \otimes E(A)$ and $Q \otimes A$ is a finitely generated $Q \otimes E(A)$-module with $(Q \otimes I)(Q \otimes A) = Q \otimes A$. By Lemma 5.8, there is $y \in Q \otimes I$ with $(1+y)(Q \otimes A) = 0$. Thus $1 + y = 0$ and $y = -1 = m/n \otimes x$ for some $m/n \in Q$, $x \in I$. Thus, $-n = m \otimes x = 1 \otimes mx = mx$ so that $n \in I$ and $nE(A) \subseteq I \subseteq E(A)$.

Now $E' = E(A)/nE(A)$ is a commutative ring, $I' = I/nE(A)$ is an ideal of E', $A' = A/nA$ is a finitely generated E'-module and $I'A' = A'$. Again, by Lemma 5.8, there is $y + nE(A) \in I'$ with $(1+y+nE(A))(A') = 0$. Thus, $(1+y)(A) \subseteq nA$ so that $1 + y \in nE(A)$ and $1 \in -y + nE(A) \subseteq I + nE(A) = I$. Consequently, $I = E(A)$.

(b) Assume that I is a right ideal of $E(A)$ with $IA = A$. Since I is $E(A)$-projective, $I \simeq H_AT_A(I)$ (Theorem 5.1). Moreover, $\mu : T_A(I) \to IA = A$, given by $\mu(r \otimes a) = ra$ is an isomorphism (Exercise 5.5). Thus, $I \simeq H_AT_A(I) \simeq H_A(A) \simeq E(A)$ and $I = fE(A)$ with $IA = A$. As in the proof of Corollary 5.7, $I = E(A)$. ///

Example 5.10. Suppose that $A = B \oplus C$ is torsion free where rank $(B) =$ rank $(C) = 1$ and type$(B) <$ type(C). Then A is not faithful.

Proof. Let $I = \{f \in E(A) \mid f(C) = 0\}$ a right ideal of $E(A)$, noting that since type$(B) <$ type(C), C is fully invariant in A. Clearly $I \neq E(A)$, yet $IA = A$ since type$(B) <$ type(C) implies $A = B \oplus C = B \oplus \mathrm{Hom}(B,C)B \subseteq IA$. ///

Assume that $G = \oplus_I A$ is a direct sum of a rank-1 torsion free group A with index set I. If C is a pure subgroup of G then $C \simeq \oplus_J A$ for some $J \subseteq I$ (Baer [2]). If C is a subgroup of G and if C is homogeneous with type(C) = type(A), i.e., $S_A(C) = C$, then $C \simeq \oplus_J A$ for some $J \subseteq I$ (Kolettis [1]). Both of these results are consequences of:

Theorem 5.11. Assume that A is torsion free of finite rank. Then the following are equivalent.

(a) A is faithful and if $B \subseteq \oplus_\Lambda A$ with $S_A(B) = B$ then $B \simeq \oplus_J A$ for some $J \subseteq \Lambda$;

(b) If $0 \neq I$ is a right ideal of $E(A)$ then I is isomorphic to $E(A)$ as $E(A)$-modules, i.e., $E(A)$ has no zero divisors and $E(A)$ is right principal.

Proof. (b) => (a) Corollary 5.7 and Theorem 5.6 implies that A is faithful. Consequently, if M is a free $E(A)$-module and N is a submodule of M then N is a free $E(A)$-module. (Exercise 5.2).

Assume that $B \subseteq \oplus_\Lambda A$ with $S_A(B) = B$. Then $H_A(B) \subseteq H_A(\oplus_\Lambda A)$, a free $E(A)$-module, so that $H_A(B)$ is a free $E(A)$-module. Consequently, $0 \to T_A H_A(B) \to T_A H_A(\oplus_\Lambda A)$ is exact and $\theta_B : T_A H_A(B) \to B$ must be an isomorphism (Exercise 5.5). Finally $B \simeq T_A H_A(B) \simeq T_A(\oplus_J E(A)) \simeq \oplus_J A$ for some $J \subseteq \Lambda$.

(a) => (b) Let $0 \neq I$ be a right ideal of $E(A)$. Then $IA \subseteq A$ with $S_A(IA) = IA$ so that $IA \simeq A$ (by (a)). Choose a monomorphism $f \in E(A)$ with $f(A) = IA$. Then $A = f^{-1}IA$ so $f^{-1}I = E(A)$, since A is faithful. Therefore, $I = fE(A) \simeq E(A)$. ///

The following is a generalization of Exercise 1.2.

Corollary 5.12: If B is a pure subgroup of A^m for some $1 \leq m \in Z$, where A is a finite rank torsion free group and $E(A)$ is a principal ideal domain, then B is a summand of A^m and $B \simeq A^n$ for some $1 \leq n \leq m$.

Proof. There is an exact sequence $0 \to H_A(B) \to H_A(A^m) \to H_A(A^m/B)$ and $H_A(A^m)$ is a free $E(A)$-module. Let $f \in H_A(A^m)$ and $I = \{g \in E(A) \mid fg \in H_A(B)\}$, a right ideal of $E(A)$. If $0 \neq I$ then $QI = QE(A)$, a field. Thus $nE(A) \subseteq I$ for some $0 \neq n \in Z$ so that $nf(A) \subseteq fI(A) \subseteq B$ whence $f(A) \subseteq B$. Therefore, $H_A(A^m)/H_A(B)$ is a finitely generated torsion free $E(A)$-module and $H_A(B)$ is a free summand of $H_A(A^m)$ since $E(A)$ is a principal ideal domain. By Theorem 5.1, $B \simeq A^n$ is a summand of A^m. ///

A torsion free group G is homogeneous separable if each finite subset of G is contained in a homogeneous completely decomposable summand of G. If A is a finite rank torsion free group then G is A-free if $G \simeq \oplus_I A$ for some I and locally A-free if every finite subset of G is contained in an A-free summand of G. Note that if $\text{rank}(A) = 1$ then G is homogeneous completely decomposable iff G is A-free. Also G is homogeneous separable iff G is locally A-free. The following theorem generalizes classical results proved for homogeneous separable groups (Fuchs [7], Vol. II, §87).

Theorem 5.13 (Arnold-Murley [1]): Suppose that A is a finite rank torsion free group and that $E(A)$ is a principal ideal domain.

(a) If G is A-free and if B is a finite rank pure subgroup of G with $S_A(B) = B$ then B is an A-free summand of G.

(b) G is locally A-free iff $S_A(G) = G$ and G is isomorphic to a pure subgroup of $S_A(\Pi_I A)$ for some index set I.

(c) If G is locally A-free and if B is a pure subgroup of G with $S_A(B) = B$ then B is locally A-free.

(d) Countable locally A-free groups are A-free.

Proof. An outline of the proof is given below.

An $E(A)$-module M is locally $E(A)$-free if each finite subset of M is contained in a finitely generated $E(A)$-free summand of M. Then H_A : locally A-free groups $\to$ locally $E(A)$-free right $E(A)$-modules is an equivalence with inverse T_A, where H_A and T_A are as defined in Theorem 5.1. This duality is then applied to analogous results of Chase [1], on locally free modules over principal ideal domains, to prove the theorem. ///

Corollary 5.14: Assume that A is finite rank torsion free and that $E(A)$ is a principal ideal domain.

(a) If $B \oplus B' \simeq \oplus_I A$ then $B \simeq \oplus_J A$ for some $J \subseteq I$.

(b) If $B \subseteq G$, $G/B \simeq A$, and $S_A(G) + B = G$ then B is a summand of G.

(c) If $B \subseteq \oplus_I A$ with $S_A(B) = B$ then $B \simeq \oplus_J A$ for some $J \subseteq I$.

(d) If B is a pure subgroup of $\oplus_I A$ with $\operatorname{rank}(B) < \infty$ and $S_A(B) = B$ then B is a summand of $\oplus_I A$ and $B \simeq A^n$ for some $0 < n \in Z$.

Proof.

(a) Theorem 5.1

(b) Corollary 5.7

(c) Theorem 5.11 and Theorem 5.9

(d) Theorem 5.13 a. and (a) . ///

The next theorem demonstrates that the hypothesis $S_A(B) = B$ is, in general, necessary for Corollary 5.14.d.

Theorem 5.15: Assume that A is reduced and torsion free of finite rank. Then every pure subgroup of A is a summand of A iff A is homogeneous completely decomposable.

Proof. ($\Leftarrow$) Corollary 5.12.

($\Rightarrow$) If every pure subgroup of A is a summand of A then it follows that $A = B_1 \oplus B_2 \oplus \ldots \oplus B_k$ with $\mathrm{rank}(B_i) = 1$ for each i. Suppose that A is not homogeneous, say $\mathrm{type}(B_1) \neq \mathrm{type}(B_2)$. Further assume that $\mathrm{type}(B_1)$ and $\mathrm{type}(B_2)$ are not comparable. Choose $0 \neq b_1 \in B_1$, $0 \neq b_2 \in B_2$ and let $C = \langle b_1 + b_2 \rangle_*$. Then C is a summand of A so that $B_1 \oplus B_2 = C \oplus D$ which is a contradiction, since $\mathrm{typeset}(B_1 \oplus B_2) = \{\mathrm{type}(C), \mathrm{type}(B_1), \mathrm{type}(B_2)\}$, $b_i = c_i + d_i \in C \oplus D$, $\mathrm{type}(C) = IT(B_1 \oplus B_2)$ and $\mathrm{type}(B_1)$, $\mathrm{type}(B_2)$ are incomparable.

Now suppose that $\mathrm{type}(B_1) < \mathrm{type}(B_2)$ and choose $0 \neq b_i \in B_i$ and $n \in Z$ such that $nx = b_2$ has no solution in A (since A is reduced). Define $C = \langle nb_1 + b_2 \rangle_*$ so that $B_1 \oplus B_2 = C \oplus D$. Now $\mathrm{type}(C) = \mathrm{type}(B_1) = \inf\{\mathrm{type}(B_1), \mathrm{type}(B_2)\}$ so that $\mathrm{type}(D) = \mathrm{type}(B_2)$ since $\mathrm{typeset}(B_1 \oplus B_2)$ has two elements in this case. If $\tau = \mathrm{type}(D)$ then $D = (B_1 \oplus B_2)(\tau) = B_2$ so $B_1 \oplus B_2 = C \oplus B_2$. But $b_1 \notin C \oplus B_2$ since $b_1 = q(nb_1 + b_2) + tb_2$ for some $q, t \in Q$ implies that $qn = 1$, $q + t = 0$, $-t = q = 1/n$, and $-tb_2 = (1/n)b_2 \in A$ which contradicts the choice of n. Thus $\mathrm{type}(B_1) < \mathrm{type}(B_2)$ is impossible. Similarly $\mathrm{type}(B_2) < \mathrm{type}(B_1)$ is impossible so that A must be homogeneous.

EXERCISES

5.1 Prove that if $G = B \oplus C$ then $S_A(G) = S_A(B) \oplus S_A(C)$ and that $S_A(S_A(G)) = S_A(G)$.

5.2 Assume that R is a right principal ring with no zero divisors. Prove that if M is a free right R-module and if N is a submodule of M then N is a free R-module. (e.g. Rotman [3]).

5.3 Characterize all faithful torsion free groups of rank 2 (see Section 2).

5.4 Give an example of a finite rank torsion free group A and an exact sequence of torsion free groups $0 \to B \to G \to A \to 0$ such that $S_A(G) = G$ and B is not a summand of G.

5.5 Assume that A is a torsion free group of finite rank and that $E(A)$ is right hereditary. (a) Prove that if I is a right ideal of $E(A)$ then $\mu : T_A(I) \to IA$, defined by $\mu(R \otimes a) = ra$ is an isomorphism. (Hint: Reduce to the case that I is finitely generated and write $I \oplus I' = F$, a finitely generated free $E(A)$-module).

(b) Prove that if M is a projective right $E(A)$-module and N is a submodule of M then N is projective. Furthermore $0 \to T_A(N) \to T_A(M)$ is exact (e.g. Rotman [3]).

5.6 Assume that A is a torsion free group of finite rank. Prove that $E(A)$ is right hereditary iff whenever $G \simeq \oplus_I A$ and B is a subgroup of G with $S_A(B) = B$ then $B \oplus B' \simeq \oplus_J A$ for some B' and some index set J.

5.7 Prove that every pure subgroup of a torsion free group A is a summand of A iff $A = B \oplus D$ where B is a finite rank homogeneous completely decomposable group and D is a finite rank divisible group.

§6. Completely Decomposable Groups and Generalizations:

The Baer-Kulikov-Kaplansky theorem states that summands of a completely decomposable group are completely decomposable. This section is devoted to direct sum decompositions of a finite rank torsion free group $G = A_1 \oplus \ldots \oplus A_n$, where each $E(A_i)$ is a principal ideal domain. More general results are proved in Arnold-Hunter-Richman [1], some of which are outlined in Exercise 6.2.

The duality of §5 does not seem relevant in this case since the choice of a group A with suitable endomorphism ring such that $G \simeq T_A H_A(G)$ is unclear. On the other hand if $A \simeq A_i$ for each i then $G \simeq A^n$ with $E(A)$ a principal ideal domain so that any summand of G is isomorphic to A^m for some m (Corollary 5.2).

A family $\{A_i\}$ of abelian groups is <u>semi-rigid</u> if whenever $i \neq j$ then $\mathrm{Hom}(A_i, A_j) = 0$ or $\mathrm{Hom}(A_j, A_i) = 0$. Charles [1], uses semi-rigid families to give a proof of the Baer-Kulikov-Kaplansky theorem. More generally, the family $\{A_i\}$ is <u>pseudo-rigid</u> if whenever $i \neq j$, $f : A_i \to A_j$, and $g : A_j \to A_i$ then $fg = 0$. The next series of results demonstrates that pseudo-rigidity is also a useful concept.

Pseudo-rigid families are constructed by using the notion of quasi-isomorphism, due to Jónsson [1]. Two finite rank torsion free groups A and B are <u>quasi-isomorphic</u> if there are monomorphisms $f : A \to B$ and $g : B \to A$ such that $B/f(A)$ and $A/g(B)$ are bounded, hence finite. Quasi-isomorphism is an equivalence relation and if A and B are isomorphic then A and B are quasi-isomorphic. Moreover, if A and B are quasi-isomorphic and $\mathrm{rank}(A) = 1$ then A and B are isomorphic (Corollary 1.3). In general, however, quasi-isomorphism does not imply isomorphism (Example 2.2).

Proposition 6.1: Suppose that A is a torsion free group of finite rank.

(a) If $h \in E(A)$ is a monomorphism then $A/h(A)$ is finite.

(b) A non-zero endomorphism f of A is a monomorphism iff f is not a zero divisor of $E(A)$.

(c) Every $0 \neq f \in E(A)$ is a monomorphism iff $QE(A)$ is a division algebra where $QE(A) = Q \otimes_Z E(A)$.

Proof. (a) There is a descending chain of right ideals $QE(A) \supseteq hQE(A) \supseteq \ldots \supseteq h^k QE(A) \supseteq \ldots$. Since $\dim_Q(QE(A))$ is finite, $h^{k+1}QE(A) = h^k QE(A)$ for some k. Write $h^k = h^{k+1}(m/n)g$ for some $m/n \in Q$, $g \in E(A)$. Then $nh^k = mh^{k+1}g$ and $h^k(mhg-n) = 0$. Since h is a monomorphism, $mhg = n$ so that $nA = hmg(A) \subseteq h(A) \subseteq A$ and $A/h(A)$ is finite (Theorem 0.1).

(b) (=>) follows from (a).

(<=) is as in the proof of (a).

(c) Note that if $f \in E(A)$ then f is a zero-divisor of $E(A)$ iff f is a zero-divisor of $QE(A)$. Thus (<=) follows from (b) while (=>) is a consequence of the fact that if $QE(A)$ is an artinian algebra with no zero-divisors then $QE(A)$ is a division algebra (Exercise 9.1). ///

Corollary 6.2: Let A and B be torsion free groups of finite rank. Then the following are equivalent:

(a) A and B are quasi-isomorphic;

(b) There is a monomorphism $f : A \to B$, such that $B/f(A)$ is finite;

(c) There is $0 \neq n \in Z$ and $f : A \to B$, $g : B \to A$ with $gf = n1_A$ and $fg = n1_B$.

(d) There are monomorphisms $f : A \to B$ and $g : B \to A$.

Proof. (a) => (b) Clear

(b) => (c) If $0 \neq n \in Z$ with $nB \subseteq f(A) \subseteq B$ let $g = f^{-1}n : B \to A$.

(c) => (d) Clearly f and g must be monomorphisms.

(d) => (a) is a consequence of Proposition 6.1(a). ///

Corollary 6.3: Let $\{A_i\}$ be a family of finite rank torsion free groups such that $E(A_i)$ has no zero divisors for each i. Then $\{A_i\}$ is a pseudo-rigid family iff A_i and A_j are not quasi-isomorphic whenever $i \neq j$.

Proof. ($\Rightarrow$) Apply Corollary 6.2.

($\Leftarrow$) Apply Proposition 6.1 and Corollary 6.2. ///

Example 6.4: There are finite rank torsion free groups A and B such that A and B are not quasi-isomorphic and $\{A,B\}$ is not a pseudo-rigid family.

Proof. As constructed in Example 2.5, there is an exact sequence $0 \to B \to A \to C \to 0$ of torsion free groups such that A is strongly indecomposable of rank 2 and B, C are rank 1 groups with $\text{type}(C) < \text{type}(B)$. Then A and B are not quasi-isomorphic and $\{A,B\}$ is not a pseudo-rigid family since the composite $A \to C \to B \to A$ need not be zero. Of course, $E(A)$ has zero divisors. ///

Corollary 6.5: Suppose that $A = A_1 \oplus \ldots \oplus A_n$, where each A_i is torsion free of finite rank and $E(A_i)$ has no zero divisors for each i. Let $\{A_j \mid j \in J\}$ be a complete set of representatives for the set of quasi-isomorphism classes of $\{A_i\}$ and let $G_j = \oplus_i\{A_i \mid A_i$ and A_j are quasi-isomorphic$\}$ for each $j \in J$. Then $A = \oplus\{G_j \mid j \in J\}$ and $\{G_j\}$ is a pseudo-rigid family.

Proof. A consequence of Corollary 6.3. ///

The following theorem is the finite rank analog of Theorem 2.13, Charles [1], using pseudo-rigidity in place of semi-rigidity. The proof makes extensive use of idempotents in endomorphism rings and, therefore, is valid in an additive category such that idempotent morphisms have kernels (see Arnold-Hunter-Richman [1]). If $G = Y \oplus X$ is a group then there are associated orthogonal idempotents $x, y \in E(G)$ such that $1_G = y + x$. If W is a subgroup of G and if $x : W \to X$ is an isomorphism then $G = Y \oplus x^{-1}(X) = Y \oplus W$. Conversely, if $G = Y \oplus W$ with associated orthogonal idempotents y' and w then $x : W \to X$ is an isomorphism with inverse $w : X \to W$ since $x = x^2 = x \cdot 1 \cdot x = x(y'+w)x = xwx$ and $w = w^2 = w(y+x)w = wxw$.

Theorem 6.6: Suppose that $G = A \oplus B = G_1 \oplus \ldots \oplus G_m$ is finite rank torsion free and that $\{G_i\}$ is a pseudo-rigid family. There are subgroups $A = A_0 \supseteq A_1 \supseteq \ldots \supseteq A_m = 0$ of A such that for each $1 \le n \le m$, $G = G_1 \oplus \ldots \oplus G_n \oplus A_n \oplus B_n$ where $A_{n-1} = A_n \oplus A'_n$ and either $A'_n = 0$ or else A'_n is isomorphic to a summand of G_n. Furthermore, $A = A'_1 \oplus \ldots \oplus A'_n$.

Proof. Let $A_{-1} = A$, $A'_0 = 0$, $B_0 = B$, and $G_0 = 0$. Then $G = A_0 \oplus B_0$ and $A_{-1} = A_0 \oplus A'_0$.

Now assume that $G = G_1 \oplus \ldots \oplus G_n \oplus A_n \oplus B_n$; $A = A_0 \supseteq \ldots \supseteq A_n$; $A_{i-1} = A_i \oplus A'_i$ for $0 \le i \le n < m$ and that $A'_i = 0$ or isomorphic to a summand of G_i for $1 \le i \le n$. By induction on n it suffices to find A_{n+1} with $A_n = A_{n+1} \oplus A'_{n+1}$, A'_{n+1} isomorphic to a summand of G_{n+1}, and B_{n+1} with $G = G_1 \oplus \ldots \oplus G_{n+1} \oplus A_{n+1} \oplus B_{n+1}$. Now $G = (G_1 \oplus \ldots \oplus G_n) \oplus G_{n+1} \oplus D_{n+1} = (G_1 \oplus \ldots \oplus G_n) \oplus A_n \oplus B_n$ with associated orthogonal idempotents $1_a = x + g_{n+1} + d_{n+1} = x' + a_n + b_n$ where $D_{n+1} = G_{n+2} \oplus \ldots \oplus G_m$ so that $a_n + b_n : G_{n+1} \oplus D_{n+1} \to A_n \oplus B_n$ is an isomorphism with inverse $g_{n+1} + d_{n+1} : A_n \oplus B_n \to G_{n+1} \oplus D_{n+1}$ (remarks preceding Theorem 6.6).

Let $H = (a_n + b_n)(G_{n+1})$ and $K = (a_n + b_n)(D_{n+1})$ so that $A_n \oplus B_n = H \oplus K$ is a summand of G where $1_{A_n \oplus B_n} = h + k = a_n + b_n$ are the associated orthogonal idempotent decompositions in $E(A_n \oplus B_n) \subseteq E(G)$.

Note that: (i) $h = ha_nh + hb_nh$.

(ii) $ha_nh = ha_nha_nh$: since $ha_nh = ha_n \cdot 1 \cdot a_nh = ha_n(h+k)a_nh = ha_nha_nh + ha_nka_nh$ and $ha_nka_nh = 0$, observing that $ha_nka_nh : H \to A_n \to K \to A_n \to H$, $H \simeq G_{n+1}$, $K \simeq D_{n+1} = G_{n+2} \oplus \ldots \oplus G_m$, and $\{G_i\}$ is a pseudo-rigid family.

(iii) ha_nh and hb_nh are idempotents.

(iv) If $e = a_n h a_n h a_n$ and $f = b_n h b_n h b_n$ then e and f are orthogonal idempotents.

(v) Define $s = h(e+f)$. Then $hs = s$, $sh = h$, and $s^2 = s$.

Now define $A_{n+1} = (a_n - e)(A_n)$ so that $A_n = A_{n+1} \oplus A'_{n+1}$ where $A'_{n+1} = e(A_n)$. Then A'_{n+1} is isomorphic to a summand of H, hence of $G_{n+1} \simeq H$, since $h : A'_{n+1} \to H$, $e : H \to A'_{n+1}$ and $ehe = e$ as a consequence of (iii).

But $s^2 = s \in E(G)$ so $G = s(G) \oplus (1-s)(G) = H \oplus (1-s)(G)$ noting that $H = s(G)$ by (v). Furthermore, $G_1 \oplus \ldots \oplus G_n \oplus A_{n+1}$ is a summand of G contained in $(1-s)(G) = \text{Kernel}(s)$, recalling that $G = G_1 \oplus \ldots \oplus G_n \oplus A_n \oplus B_n$ so that $s(G_1 \oplus \ldots \oplus G_n) = 0$ and that $s(a_n - e) = h(e+f)(a_n - e) = hea_n - he = he - he = 0$. Consequently, $(1-s)(G) = G_1 \oplus \ldots \oplus G_n \oplus A_{n+1} \oplus B_{n+1}$ for some B_{n+1} (not necessarily contained in B_n).

Therefore, $G = (1-s)(G) \oplus H = G_1 \oplus \ldots \oplus G_n \oplus A_{n+1} \oplus B_{n+1} \oplus H$. Finally, $s : G_{n+1} \to H = s(G) = s(A_n \oplus B_n)$ is an isomorphism since $a_n + b_n : G_{n+1} \to H$ is an isomorphism, $s(a_n + b_n) = s$, and $sh = h$. Thus $G = G_1 \oplus \ldots \oplus G_n \oplus A_{n+1} \oplus B_{n+1} \oplus G_{n+1}$, by the remarks preceding Theorem 6.6, as desired. ///

Corollary 6.7. Assume that $G = A \oplus B = C_1 \oplus \ldots \oplus C_m$ is finite rank torsion free where each $E(C_i)$ has no zero divisors. Then $A = A_1 \oplus \ldots \oplus A_n$ where for each i, A_i is isomorphic to a summand of some $G_j = \oplus \{C_k | C_k$ is quasi-isomorphic to $C_j\}$.

Proof. Apply Corollary 6.5 and Theorem 6.6. ///

Corollary 6.8: If $G = A \oplus B = C_1 \oplus \ldots \oplus C_m$ is completely decomposable, where each C_i is torsion free of rank 1 then $A \simeq \oplus \{C_i | i \in I\}$ for some subset I of m^+.

Proof. Apply Corollary 6.7 to reduce to the case that G is homogeneous completely decomposable, observing that quasi-isomorphism implies isomorphism for rank-1 groups. Now apply Corollary 5.3. ///

The next example shows that a group $G = C_1 \oplus \ldots \oplus C_m$ with each $E(C_i)$ a principal ideal domain and C_i quasi-isomorphic to C_j for each i and j may have non-equivalent direct sum decompositions.

Example 6.9: There are finite rank torsion free groups A_1, A_2, A_3, A_4 such that $A_1 \oplus A_2 = A_3 \oplus A_4$, A_i is quasi-isomorphic to A_j for each i and j, $E(A_i) \simeq Z$ for each i, and A_4 is not isomorphic to A_1 or A_2.

Proof. Let X_1 and X_2 be subgroups of Q such that type(X_1) and type(X_2) are incomparable; $1 \in X_i$; for some prime p, $h_p^{X_i}(1) = 0$; and $E(X_i) \simeq Z$ (e.g. type$(X_i) = [(k_{iq})]$ where $k_{iq} < \infty$ for each q). Define $A = \langle X_1a_1, X_2a_2, (a_1+a_2)/p^{\infty}\rangle \subseteq Qa_1 \oplus Qa_2$. Then $E(A) \simeq Z$ (as in Example 2.4). If $p \neq q$ is a prime then q-rank$(A) = 2$, since $A/qA \simeq Z/qZ \oplus Z/qZ$ whenever $q \neq p$.

Let $p_i \neq p$, $i = 1, 2$ be distinct primes. Choose $a_i \in A\backslash p_iA$ and define $A_i = E(A)a_i + p_iA = Za_i + p_iA$. Then A is not isomorphic to A_i since A/p_iA is not cyclic and $E(A) \simeq Z$. Moreover, $E(A_i) \simeq E(A) \simeq Z$ for $i = 1, 2$, since A/A_i is finite. Write $1 = rp_1 + sp_2$ for some r, $s \in Z$. Then $p_1 \oplus p_2 : A \to A_1 \oplus A_2$ and $r \oplus s : A_1 \oplus A_2 \to A$ with $rp_1 + sp_2 = 1$ so that $A \oplus A_3 \simeq A_1 \oplus A_2$ where $A_3 \simeq r(A_1) \cap s(A_2)$ is quasi-isomorphic to A. Thus, $E(A) \simeq Z$ and letting $A = A_4$ completes the proof. ///

If $C_1, \ldots, C_m$ are finite rank torsion free groups with C_i quasi-isomorphic to C_j for each i and j and if each $E(C_i)$ is a principal ideal domain then there is a principal ideal domain $R \simeq E(C_i)$ for each i such that each C_i is an R-module; $\mathrm{Hom}_R(C_i,C_j) = \mathrm{Hom}_Z(C_i,C_j)$; $rh = hr$ for each $r \in R$, $h \in \mathrm{Hom}(C_i,C_j)$; and if A is a group summand of $C_1 \oplus \ldots \oplus C_m$ then A is an R-module summand (Exercise 6.4).

Henceforth, R is identified with $E(C_i)$ for each i.

Lemma 6.10: Suppose that C_1 and C_2 are quasi-isomorphic torsion free groups of finite rank and that $R = E(C_1) = E(C_2)$ is a principal ideal domain.

(a) If $f \in \mathrm{Hom}(C_1, C_2)$, $g \in \mathrm{Hom}(C_2, C_1)$ with $fg = gf = uv \in R$ for some $u, v \in R$ with $uR + vR = R$ then $C_1 \oplus C_2 = C \oplus D$ for some C quasi-isomorphic to C_1 and C_2 with $\alpha : C \to C_1$, $\alpha' : C_1 \to C$, $\beta : C \to C_2$, $\beta' : C_2 \to C$ such that $\alpha\alpha' = u^2$, $\beta\beta' = v$ and $E(C) \simeq R$. Moreover, $\alpha'\alpha = u^2 \in R$ and $\beta'\beta = v \in R$.

(b) If A is finite rank torsion free with $\mu_1, \mu_2 : C_1 \to A$, $\lambda_1, \lambda_2 : A \to C_1$ and $\lambda_1\mu_1 R + \lambda_2\mu_2 R = rR$ then there is $\mu : C_1 \to A$ and $\lambda : A \to C_1$ with $\lambda\mu = r$.

Proof. (a) Write $1 = su^3 + tv$ for some $s, t \in R$, noting that $R = u^3R + vR$ since $R = uR + vR$. Let $e \in E(C_1 \oplus C_2)$ be represented by $\begin{pmatrix} su^3 & tug \\ suf & tv \end{pmatrix}$. Then $e^2 = e$.

Write $G = C_1 \oplus C_2 = C \oplus D$ where $C = e(G)$ and let $1 = c_1 + c_2$ be orthogonal idempotents associated with $G = C_1 \oplus C_2$. Define

$$\alpha' = \begin{pmatrix} u^2 & 0 \\ f & 0 \end{pmatrix} \text{ and } \beta' = \begin{pmatrix} 0 & ug \\ 0 & v \end{pmatrix}.$$

Then $\alpha' : C_1 \to C$ since $e\alpha' = \alpha' = \alpha' c_1$, and $\beta' : C_2 \to C$ since $e\beta' = \beta' = \beta' c_2$. Thus $\alpha = c_1 e : C \to C_1$ and $\beta = c_2 e : C \to C_2$ with $\alpha\alpha' = c_1 e\alpha' = u^2$ and $\beta\beta' = c_2 e\beta' = c_2\beta' = v$. Also $\alpha'\alpha e = \alpha' c_1 e = \alpha' e = u^2 e$ and $\beta'\beta e = ve$ so that $\alpha'\alpha = u^2 \in R$ and $\beta\beta' = v \in R \subseteq E(C)$. Finally, C is quasi-isomorphic to C_1 and C_2 so that $R = E(C)$ since R is a principal ideal domain and $E(C)/R$ is finite.

(b) Let $M = \begin{pmatrix} \lambda_1\mu_1 & \lambda_1\mu_2 \\ \lambda_2\mu_1 & \lambda_2\mu_2 \end{pmatrix} \in \mathrm{Mat}_2(R)$. Since R is a principal ideal domain there are invertible matrices L and K in $\mathrm{Mat}_2(R)$ with $LMK = \begin{pmatrix} d & o \\ o & x \end{pmatrix}$ for some $x \in R$ where $d = \text{g.c.d.}(\lambda_1\mu_1, \lambda_1\mu_2, \lambda_2\mu_1, \lambda_2\mu_2)$ (Exercise 6.3). If (s,t) is the first row of L and $\begin{pmatrix} u \\ v \end{pmatrix}$ is the first column of K then $d = (s\lambda_1 + t\lambda_2)(\mu_1 u + \mu_2 v) \in E(C_1) = R$. Write $r = r'd$ and let $\mu = \mu_1 u + \mu_2 v$, $\lambda = r's\lambda_1 + r't\lambda_2$. ///

Theorem 6.11: Suppose that $G = A \oplus B = C_1 \oplus \ldots \oplus C_m$ is finite rank torsion free where each $E(C_i)$ is a principal ideal domain. Then $A = A_1 \oplus \ldots \oplus A_n$ where for each i, A_i is quasi-isomorphic to some C_j and $E(A_i) \simeq E(C_j)$.

Proof. By Corollary 6.7 and the remarks preceding Lemma 6.10, it is sufficient to assume that each C_i and C_j are quasi-isomorphic and $R = E(C_i)$ for each i.

It is sufficient to prove that $G = A \oplus B = C \oplus D$ for some C quasi-isomorphic to each C_i with $R = E(C)$ and with homomorphisms $\mu : C \to A$, $\lambda : A \to C$, $\mu' : C \to A$, $\lambda' : A \to C$ such that $\lambda\mu R + \lambda'\mu' R = R$, in which case the proof is complete by Lemma 6.10.b and an induction on the rank of A.

Let $1_G = a + b = c_1 + \ldots + c_m$ be the associated idempotent projections of $G = A \oplus B = C_1 \oplus \ldots \oplus C_m$ where $0 \neq a$. For each i, $c_i a c_i + c_i b c_i = c_i$. If $c_i a c_i = 0$ for each i then $B \simeq C_1 \oplus \ldots \oplus C_m$ and $A = 0$, a contradiction. For some i, say $i = 1$, $c_1 a c_1 \neq 0$. Thus there is $\mu_1 = ac_1 : C_1 \to A$ and $\lambda_1 = c_1 a : A \to C_1$ with $0 \neq \lambda_1\mu_1 \in R$. If $\lambda_1\mu_1 R = R$ then C_1 is isomorphic to a summand of A and the proof is complete.

Now assume that $\lambda_1\mu_1 R = P_1^{e_1} \ldots P_k^{e_k}$ is a product of powers of distinct prime ideals P_i of R. Note that $a = \Sigma\{c_i a c_j \mid 1 \leq i, j \leq m\}$ and $a^m = \Sigma\{c_{i_1} a c_{i_2} \ldots a c_{i_{m+1}} \mid \{i_1, \ldots, i_{m+1}\} \subseteq m^+\} = a$. Hence each term has a repeated subscript so there is some $j > 1$, say $j = 2$, and $\mu_2 : C_2 \to A$, $\lambda_2 : A \to C_2$ with $\lambda_2\mu_2 \notin P_k$, otherwise $a \in P_k$, $E(A) = P_k E(A)$ and $R = P_k R$ (since $E(G) \simeq \mathrm{Mat}_m(R)$) contradicting the assumption that P_k is a prime ideal of the principal ideal domain R.

Since C_1 and C_2 are quasi-isomorphic there is $f : C_1 \to C_2$, $g : C_2 \to C_1$ with $fg \neq 0$. Write $fg = uv \in R$ where $vR \not\subseteq P_i$ for each i and uR is a product of powers of the P_i's so that $uR + vR = R$. By Lemma 6.10.a, $C_1 \oplus C_2 = C \oplus D$ for some C quasi-isomorphic to C_1 and C_2 with $E(C) = R$ and $\alpha : C \to C_1$, $\alpha' : C_1 \to C$, $\beta : C \to C_2$ and $\beta' : C_2 \to C$ with $\alpha\alpha' = u^2 = \alpha'\alpha$ and $\beta\beta' = v = \beta'\beta$.

Now $\mu = \mu_1\alpha : C \to A$ and $\lambda = \alpha'\lambda_1 : A \to C$ with $\lambda\mu = \alpha'\lambda_1\mu_1\alpha = \lambda_1\mu_1 u^2$ and $\mu' = \mu_2\beta : C \to A$, $\lambda' = \beta'\lambda_2 : A \to C$ with $\lambda'\mu' = \beta'\lambda_2\mu_2\beta = \lambda_2\mu_2\beta'\beta = \lambda_2\mu_2 v$. Thus $\lambda\mu R + \lambda'\mu' R = P_1^{f_1} \ldots P_{k-1}^{f_{k-1}}$ for some f_i since $\lambda_2\mu_2 \not\subseteq P_k$, uR and $\lambda_1\mu_1 R$ are products of powers of $P_1, \ldots, P_k$, and $vR \not\subseteq P_i$ for any i. By Lemma 6.10.b there is $\mu_0 : C \to A$ and $\lambda_0 : A \to C$ with $\lambda\mu R + \lambda'\mu' R = \lambda_0\mu_0 R$. By induction on k, the proof is complete. ///

As a consequence of Theorem 6.11, Example 6.9 is a prototype for non-equivalent direct sum decompositions of a finite rank torsion free group $G = C_1 \oplus \ldots \oplus C_m$ where each $E(C_i)$ is a principal ideal domain.

EXERCISES

<u>6.1</u> Let A be a torsion free group of finite rank. Prove that the following are equivalent for $f \in E(A)$.

(a) f is not a zero divisor of $E(A)$;

(b) f is a monomorphism;

(c) There is $g \in E(A)$ and $0 \neq n \in Z$ with $fg = n = gf$;

(d) f is a unit of $QE(A)$.

<u>6.2</u> (a) Adapt the proof of Theorem 6.6 to prove that if $G = A \oplus B = \oplus G_i$ where $\{G_i\}$ is a countable psuedo rigid family then $A = \oplus A_n'$ where each A_n' is isomorphic to a summand of G_n.

(b) Prove that if $G = A \oplus B = \oplus_I G_i$ for some index set I with each G_i countable then $A = \oplus_J A_j$ where each A_j is isomorphic to a summand of countably many of the G_i's. (Kaplansky's Theorem, Fuchs [7], Vol. I).

(c) Prove that if $G = A \oplus B = \oplus_I G_i$ for some index set I where $\{G_i | i \in I\}$ is a psuedo rigid family then $A = \oplus_J A'_j$ for some subset J of I where each A'_j is isomorphic to a summand of G_j.

(d) (Baer-Kulikov-Kaplansky) Prove that if $G = A \oplus B = \oplus_I G_i$ where each G_i is torsion free of rank 1 then $A = \oplus_J A'_j$ for some subset J of I where each A'_j is isomorphic to some G_i.

<u>6.3</u> (Invariant Factor Theorem) Suppose that R is a principal ideal domain and that $M \in \mathrm{Mat}_2(R)$. Let $M = \begin{pmatrix} a & b \\ c & d \end{pmatrix}$ and define g.c.d.(M) = g.c.d.(a,b,c,d)

(a) Show that if L is an invertible matrix in $\mathrm{Mat}_2(R)$ then g.c.d.(M) = g.c.d.(LM) = g.c.d.(ML).

(b) Prove that there are invertible matrices L and K in $\mathrm{Mat}_2(R)$ with $LM = \begin{pmatrix} e & b' \\ c' & d' \end{pmatrix}$ and $MK = \begin{pmatrix} f & b'' \\ c'' & d'' \end{pmatrix}$ where e = g.c.d.(a,c) and f = g.c.d.(a,b).

(c) Prove that there are invertible matirces L and K in $\mathrm{Mat}_2(R)$ with $LMK = \begin{pmatrix} x & o \\ o & y \end{pmatrix}$ where x = g.c.d.(M) and x divides y.

<u>6.4</u> Let A and B be quasi-isomorphic torsion free groups of finite rank such that $E(A)$ and $E(B)$ are principal ideal domains.

(a) Let $f \in \mathrm{Hom}(A,B)$ be a quasi-isomorphism and define $I_f = \{fg | g \in \mathrm{Hom}(B,A)\}$ a non-zero ideal of $E(B)$. For $\lambda \in E(A)$ define $\theta_\lambda : I_f \to I_f$ by $\theta_\lambda(fg) = f\lambda g$. Show that θ_λ is a well-defined $E(B)$-homomorphism and that θ_λ is left multiplication by some $\mu_\lambda \in E(B)$.

(b) Prove that $\phi_f : E(A) \to E(B)$ defined by $\phi_f(\lambda) = u_\lambda$ is a unique ring homomorphism with the property that $\phi_f(\lambda)fg = f\lambda g$ for each $\lambda \in E(A)$ and each $g \in \mathrm{Hom}(B,A)$.

(c) If C is a torsion free group quasi-isomorphic to B and if $g \in Hom(B,C)$ is a quasi-isomorphism then prove that $\phi_{gf} = \phi_g \phi_f : E(A) \to E(C)$.

(d) Show that if $C = A$ then $\phi_{gf} = 1_A \in E(A)$. Deduce that ϕ_f is independent of the quasi-isomorphism f.

(e) Deduce that there is a unique ring isomorphism $\phi_{BA}: E(A) \to E(B)$ such that $\phi_{BA}(\lambda)fg = f\lambda g$ for each $\lambda \in E(A)$, $g \in Hom(B,A)$, and quasi-isomorphism $f \in Hom(B,A)$. Moreover, $\phi_{CB}\phi_{BA} = \phi_{CA}$.

(f) Prove that $\phi_{BA}(\lambda)h = h\lambda$ for each $h \in Hom(A,B)$ and $\lambda \in E(A)$.

(g) Prove that if $C_1, \ldots, C_m$ are pairwise quasi-isomorphic torsion free groups of finite rank such that each $E(C_i)$ is a principal ideal domain then there is a ring $R \simeq E(C_i)$ for each i such that each C_i is an R-module, $Hom_R(C_i, C_j) = Hom_Z(C_i, C_j)$ for each i and j, $rh = hr$ for each $r \in R$ and $h \in Hom(C_i, C_j)$, and if A is a group summand of $C_1 \oplus \ldots \oplus C_m$ then A is an R-module summand.

6.5 Let p and q be distinct primes of Z and let $Z_{pq} = Z_p \cap Z_q$ where Z_p and Z_q are localizations of Z at p and q, respectively. Let $\{\alpha_i\}$ be a sequence of elements of $Z\backslash pZ$ converging to a p-adic integer α which is transcendental over Q. Define A to be the Z_{pq}-submodule of $Qx \oplus Qy$ generated by $Z_p x$, y, and $\{(x+\alpha_i y)/p^i \mid i = 0,1,2,\ldots\}$.

(a) Prove that $|\text{typeset}(A)| = 2$ and that $QE(A) \simeq Q$ (see Theorems 3.2 and 3.3).

(b) Prove that $A/Z_p x \simeq Z_q$.

(c) Let $\{\beta_i\}$ be a sequence of elements of $Z\backslash qZ$ converging to a q-adic integer β which is transcendental over Q. Define B to be the Z_{pq} submodule of $Qx \oplus Qy$ generated by $Z_q x$, y, and $\{(x+\beta_i y)/q^i \mid i = 0,1,2,\ldots\}$. Show that $\{A,B\}$ is a pseudo-rigid family but not a semi-rigid family of groups.

§7. Additive Categories, Quasi-Isomorphism and Near-Isomorphism:

A **category** is a class $\mathcal{C}$ of objects together with a set of morphisms $\mathrm{Hom}_{\mathcal{C}}(A,B)$ for each A,B in $\mathcal{C}$ and a composition $\mathrm{Hom}_{\mathcal{C}}(B,C) \times \mathrm{Hom}_{\mathcal{C}}(A,B) \to \mathrm{Hom}_{\mathcal{C}}(A,C)$, written $(g,f) \to gf$, for each A,B,C in $\mathcal{C}$ satisfying:

(i) composition is associative;

(ii) for each A in $\mathcal{C}$ there is $1_A \in \mathrm{Hom}_{\mathcal{C}}(A,A)$ with $f1_A = f$ and $1_A g = g$ whenever $f \in \mathrm{Hom}_{\mathcal{C}}(A,B)$ and $g \in \mathrm{Hom}_{\mathcal{C}}(C,A)$.

A category is additive if, in addition,

(iii) for each A,B in $\mathcal{C}$, $\mathrm{Hom}_{\mathcal{C}}(A,B)$ is an abelian group such that $g(f_1+f_2) = gf_1 + gf_2$ and $(f_1+f_2)h = f_1h + f_2h$ whenever $f_i \in \mathrm{Hom}_{\mathcal{C}}(A,B)$, $g \in \mathrm{Hom}_{\mathcal{C}}(B,C)$ and $h \in \mathrm{Hom}_{\mathcal{C}}(D,A)$; and

(iv) $\mathcal{C}$ has finite direct sums: given $A_1, \ldots, A_n$ in $\mathcal{C}$ there is an A in $\mathcal{C}$ and morphisms $i_j \in \mathrm{Hom}_{\mathcal{C}}(A_j, A)$ such that if $f_j \in \mathrm{Hom}_{\mathcal{C}}(A_j, B)$ for each $1 \le j \le n$ then there is a unique $f \in \mathrm{Hom}_{\mathcal{C}}(A,B)$ with $fi_j = f_j$ for each $1 \le j \le n$. The object A is called a direct sum of $A_1, \ldots, A_n$ and $i_1, \ldots, i_n$ are called injections for the direct sum. Denote A by $A_1 \oplus \ldots \oplus A_n$.

Lemma 7.1: Suppose that $\mathcal{C}$ is an additive category.

(a) $E_{\mathcal{C}}(A) = \mathrm{Hom}_{\mathcal{C}}(A,A)$ is a ring with identity 1_A.

(b) An object A with morphisms $i_j \in \mathrm{Hom}_{\mathcal{C}}(A_j,A)$ for $1 \le j \le n$ is a direct sum of $A_1, \ldots, A_n$ with injections $i_1, \ldots, i_n$ iff for each j there is $p_j \in \mathrm{Hom}_{\mathcal{C}}(A,A_j)$ such that $p_ji_k = 0 \in \mathrm{Hom}_{\mathcal{C}}(A_k,A_j)$ if $j \ne k$, $p_ji_k = 1_{A_j}$ if $j = k$, and $1 = i_1p_1 + \ldots + i_np_n$. In this case, i_jp_j is an idempotent of $E_{\mathcal{C}}(A)$ for each $1 \le j \le n$.

Proof. (a) is routine.

(b) (⟹) Fix j and define $\delta_{jk} \in \mathrm{Hom}\,(A_k,A_j)$ by $\delta_{jk} = 1_{A_j}$ if $k = j$ and 0 otherwise. There is $p_j \in \mathrm{Hom}_{\mathcal{C}}(A,A_j)$ with $p_ji_k = \delta_{jk}$ for each k.

Then $\Sigma\{i_jp_j | 1 \le j \le n\} \in E_{\mathcal{C}}(A)$ and $(\Sigma_j i_jp_j)i_k = \Sigma_j i_jp_ji_k = i_k = 1_A i_k$ for each k. By the uniqueness property for direct sums, $1_A = \Sigma_j i_jp_j$.

($\Leftarrow$) Given $f_j \in \mathrm{Hom}_{\mathcal{C}}(A_j, B)$ for $1 \le j \le n$, let $f = \Sigma_j f_jp_j \in \mathrm{Hom}_{\mathcal{C}}(A,B)$. Then $fi_k = \Sigma_j f_jp_ji_k = f_k$ for each k. Furthermore, if $f' \in \mathrm{Hom}_{\mathcal{C}}(A,B)$ with $f'i_k = f_k$ for each k then $f' = f'1_A = \Sigma_j f'i_jp_j = \Sigma f_jp_j = f$. Thus A is a direct sum of $A_1, \ldots, A_n$. ///

The p_j's are called <u>projections</u> for the direct sum $A \simeq A_1 \oplus \ldots \oplus A_n$ with injections $i_1, \ldots, i_n$.

An element $f \in \mathrm{Hom}_{\mathcal{C}}(A,B)$ is an <u>isomorphism</u> if there is $g \in \mathrm{Hom}_{\mathcal{C}}(B,A)$ with $fg = 1_B$ and $gf = 1_A$, in this case <u>A is isomorphic to B in $\mathcal{C}$.</u> If A and A' are direct sums of $A_1, A_2, \ldots, A_n$ in $\mathcal{C}$ then the uniqueness property implies that A and A' are isomorphic in $\mathcal{C}$. An object A in $\mathcal{C}$ is <u>indecomposable in $\mathcal{C}$</u> if whenever $A \simeq B \oplus C$ then $B = 0$ or $C = 0$.

<u>Idempotents split in the additive category $\mathcal{C}$</u> if whenever $e \in E_{\mathcal{C}}(A)$ is an idempotent then there is B in $\mathcal{C}$ and $q \in \mathrm{Hom}_{\mathcal{C}}(B,A)$, $p \in \mathrm{Hom}_{\mathcal{C}}(A,B)$ with $qp = e$ and $pq = 1_B$.

<u>Lemma 7.2</u>: Suppose that $\mathcal{C}$ is an additive category and that idempotents split in $\mathcal{C}$.

(a) If e is an idempotent of $E_{\mathcal{C}}(A)$ then there are $q_i \in \mathrm{Hom}_{\mathcal{C}}(B_i, A)$, $p_i \in \mathrm{Hom}_{\mathcal{C}}(A, B_i)$ with $p_iq_i = 1_{B_i}$, $q_1p_1 = e$, $q_2p_2 = 1_A - e$. Moreover, $A = B_1 \oplus B_2$ with injections $\{q_1, q_2\}$.

(b) A is indecomposable in $\mathcal{C}$ iff the only idempotents of $E_{\mathcal{C}}(A)$ are 0 and 1_A.

(c) If $q_1 \in \mathrm{Hom}_{\mathcal{C}}(B_1,A)$, $p_1 \in \mathrm{Hom}_{\mathcal{C}}(A,B_1)$ such that p_1q_1 is a unit of $E_{\mathcal{C}}(B_1)$ then there is $B_2 \in \mathcal{C}$ and $q_2 \in \mathrm{Hom}_{\mathcal{C}}(B_2,A)$ such that $A \simeq B_1 \oplus B_2$ with injections $\{q_1, q_2\}$.

(d) Let $i_C : C \to G$ be an inclusion and $G = A \oplus B$ with injections i_A, i_B and projections p_A, p_B. If $p_Ai_C : C \to A$ is an isomorphism then $G = C \oplus B$ with injections i_C, i_B and projections $p_C = (p_Ai_C)^{-1}p_A$, $p' = p_B(1 - i_Cp_C)$.

Proof. (a) Note that $1_A - e$ is an idempotent of $E_C(A)$. Since idempotents split there is $q_i \in Hom_C(B_i, A)$, $p_i \in Hom_C(A, B_i)$ with $p_iq_i = 1_{B_i}$, $q_1p_1 = e$, and $q_2p_2 = 1_A - e$. Then $1_A = q_1p_1 + q_2p_2$, $p_iq_i = 1_{B_i}$, $p_1q_2 = (p_1q_1p_1)(q_2p_2q_2) = (p_1)(e)(1_A-e)q_2 = 0$, and similarly $p_2q_1 = 0$. Now apply Lemma 7.1 to see that $A = B_1 \oplus B_2$ with injections $\{q_1,q_2\}$.

(b) (⟹) follows from (a); (⟸) is a consequence of Lemma 7.1.

(c) Let $p_1' = (p_1q_1)^{-1}p_1 \in Hom_C(A,B_1)$. Then $p_1'q_1 = 1_{B_1}$ and $e = q_1p_1'$ is an idempotent in $E_C(A)$. Choose $q_2 \in Hom_C(B_2,A)$, $p_2 \in Hom_C(A,B_2)$ with $q_2p_2 = 1_A - e$, $p_2q_2 = 1_{B_2}$ and apply (a).

(d) is a consequence of the proof of (c) and the fact that $1-i_Cp_C = i_Bp'$.

If C is an additive category and if the only idempotents of $E_C(A)$ are 0 and 1_A then A is indecomposable in C. An example of an indecomposable object A in an additive category C with an idempotent $e \in E_C(A)$ such that $e \neq 0$ or 1_A is given below (i.e., e does not split in C). Following is the additive category analog of the remark preceding Theorem 6.6.

Lemma 7.3. Assume that C is an additive category, that $A \simeq A_1 \oplus A_2$ with injections and projections $\{i_1,i_2\}$, $\{p_1,p_2\}$, and that $A \simeq A_1' \oplus A_2'$ with injections $\{i_1',i_2'\}$ and $\{p_1',p_2'\}$ respectively. If $p_1'i_1 : A_1 \to A_1'$ is an isomorphism then A_2 is isomorphic to A_2' in C.

Proof. As in Lemma 7.2.d, $A \simeq A_1 \oplus A_2'$ with injections $\{i_1,i_2'\}$ and projections $\{\bar{p}_1,\bar{p}_2\}$, where $\bar{p}_1 = (p_1'i_1)^{-1}p_1'$ and $\bar{p}_2 = p_2'(1_A-i_1\bar{p}_1)$. But $p_2i_2' : A_2' \to A_2$ is an isomorphism with inverse $\bar{p}_2i_2 : A_2 \to A_2'$ since $p_2i_2'\bar{p}_2i_2 = p_2(i_1\bar{p}_1 + i_2'\bar{p}_2)i_2 = p_2(1_A)i_2 = p_2i_2 = 1_{A_2}$ and $\bar{p}_2i_2p_2i_2' = \bar{p}_2(i_1p_1 + i_2p_2)i_2' = \bar{p}_2i_2' = 1_{A_2}$. ///

A ring R is local if the sum of any two non-units of R is a non-unit of R; equivalently the set M of all non-units of R is an ideal of R. In this case, M is a unique maximal ideal of R. Note that if $E_C(A)$ is local then A is indecomposable in C (Lemma 7.2.b (⟸)).

The following theorem is the Krull-Schmidt theorem for additive categories (Bass [1]).

Theorem 7.4. Suppose that $\mathcal{C}$ is an additive category such that idempotents split in $\mathcal{C}$ and that $A \simeq A_1 \oplus \ldots \oplus A_n$, where each $E_{\mathcal{C}}(A_i)$ is a local ring.

(a) If $A \simeq B_1 \oplus \ldots \oplus B_m$ then each B_j is a finite direct sum of indecomposable objects in $\mathcal{C}$.

(b) If each B_j is indecomposable in $\mathcal{C}$ then $m = n$ and there is a permutation σ of n^+ such that A_i is isomorphic to $B_{\sigma(i)}$ in $\mathcal{C}$ for each i.

Proof. Let $A \simeq A_1 \oplus \ldots \oplus A_n$ and $A \simeq B_1 \oplus \ldots \oplus B_m$ with injections and projections $\{i_1, \ldots, i_n\}$, $\{p_1, \ldots p_n\}$; $\{i_1', \ldots, i_m'\}$, $\{p_1', \ldots, p_m'\}$; respectively. Then $1_{A_1} = p_1 i_1 = p_1(\Sigma_k i_k' p_k') i_1 = \Sigma_k p_1 i_k' p_k' i_1$. Since $E_{\mathcal{C}}(A_1)$ is a local ring for some k, say $k = 1$, $(p_1 i_k')(p_k' i_1)$ is a unit of $E_{\mathcal{C}}(A_1)$. By Lemma 7.2.d, $B_1 \simeq A_1 \oplus B_1''$ with injections $\{p_1' i_1, i_1''\}$ and projections $\{\bar{p}_1, p_1''\}$, where $\bar{p}_1 = (p_1 i_1' p_1' i_1)^{-1} p_1 i_1'$. Thus $A \simeq A_1 \oplus A_2 \oplus \ldots \oplus A_n$ and $A \simeq A_1 \oplus B_1'' \oplus B_2 \oplus \ldots \oplus B_n$ with $i_1 : A_1 \to A$ as an injection of A_1 in the first sum and $\bar{p}_1 p_1'$ as a projection of A_1 in the second sum. But $(\bar{p}_1 p_1') i_1 = (p_1 i_1' p_1' i_1)^{-1} p_1 i_1' p_1' i_1 = 1_{A_1}$ so that $A_2 \oplus \ldots \oplus A_n$ is isomorphic to $B_1'' \oplus B_2 \oplus \ldots \oplus B_m$ (by Lemma 7.3).

Now (a) follows by induction on n, since A_1 is isomorphic to an indecomposable summand of B_1. Finally, (b) is true, since if B_1 is indecomposable then $B_1'' = 0$ and A_1 is isomorphic to B_1 in $\mathcal{C}$. ///

Example 7.5: If R is a ring and if $\mathcal{C}$ is a class of R-modules closed under finite direct sums and containing 0 then $\mathcal{C}$ is an additive category, where $\mathrm{Hom}_{\mathcal{C}}(A,B) = \mathrm{Hom}_R(A,B)$. Idempotents split in $\mathcal{C}$ iff $\mathcal{C}$ is closed under direct summands. In particular, if $\mathcal{A}$ is the category of finite rank torsion free abelian groups (including 0) then $\mathcal{A}$ is an additive category and idempotents split in $\mathcal{A}$.

Example 7.6: Define a category $Q\mathcal{A}$ by letting objects($Q\mathcal{A}$) = objects($\mathcal{A}$) and define $\mathrm{Hom}_{Q\mathcal{A}}(A,B) = Q \otimes_Z \mathrm{Hom}(A,B) = \mathrm{QHom}(A,B)$. Then $Q\mathcal{A}$ is an additive category and idempotents split in $Q\mathcal{A}$.

Proof. (i) Composition, $\mathrm{QHom}(B,C) \times \mathrm{QHom}(A,B) \to \mathrm{QHom}(A,C)$, defined by $(q_1g, q_2f) \to q_1q_2gf$ is associative and $1_A = 1 \otimes 1_A$ is an identity.

(ii) $\mathrm{QHom}(A,B)$ is an abelian group and composition is distributive.

(iii) $Q\mathcal{A}$ has finite direct sums; if $A_1, \ldots, A_n$ are objects of $Q\mathcal{A}$ let $A = A_1 \oplus \ldots \oplus A_n$ be a group direct sum with injections $\{i_1, \ldots, i_n\}$. Then A is a direct sum in $Q\mathcal{A}$ with injections $\{i_1, \ldots, i_n\}$ since if $f_j \in \mathrm{QHom}(A_j,B)$ then there is $0 \neq n \in Z$ with $nf_j \in \mathrm{Hom}(A_j,B)$ for each j. Thus, there is a unique $f \in \mathrm{Hom}(A,B)$ with $fi_j = nf_j$ for each j. But $(1/n)f \in \mathrm{QHom}(A,B)$ and $(1/n)fi_j = f_j$ for each j. The uniqueness of $(1/n)f$ follows from the uniqueness of f.

(iv) Idempotents split in $Q\mathcal{A}$: Let $e = (m/n)f \in \mathrm{QE}(A)$ for some $m/n \in Q$, $f \in E(A)$ be an idempotent of $\mathrm{QE}(A)$. Define $B = f(A) \in Q\mathcal{A}$ $i_B : B \to A$ the inclusion map, $q = (1/n)i_B \in \mathrm{QHom}(B,A)$, $p = mf \in \mathrm{QHom}(A,B)$. Then $qp = (m/n)i_Bf = (m/n)f = e$ and $pq = (m/n)fi_B = 1_B$ since $mnpq(b) = m^2f(b) = m^2f^2(a) = nmf(a) = nmb$ (noting that $e^2 = (m^2/n^2)f^2 = e = (m/n)f$ implies that $m^2f^2 = nmf$ and $b = f(a)$ for some $A \in \mathcal{A}$). ///

The notions of isomorphism and direct sums in $Q\mathcal{A}$ have group theoretic interpretations which, in fact, preceded the categorical concepts (see B. Jónsson [1]).

Corollary 7.7. Let A and B be torsion free groups of finite rank.

(a) A is isomorphic to B in $Q\mathcal{A}$ iff there is $f : A \to B$, $g : B \to A$ and $0 \neq n \in Z$ with $fg = n1_B$ and $gf = n1_A$.

(b) The following are equivalent for $f \in \mathrm{Hom}(A,B)$; (i) For each $0 \neq m/n \in Q$, $(m/n)f \in \mathrm{QHom}(A,B)$ is an isomorphism in $Q\mathcal{A}$; (ii) f is a monomorphism and $B/f(A)$ is finite; (iii) f is a monomorphism and there is a monomorphism $g : B \to A$.

(c) A is a direct sum of $A_1, \ldots, A_n$ in QA iff there is a monomorphism f from the group direct sum $A_1 \oplus \ldots \oplus A_n$ to A such that $A/\text{Image } f$ is finite.

(d) A is indecomposable in QA iff whenever $nA \subseteq B \oplus C \subseteq A$ with $0 \neq n \in Z$ then $B = 0$ or $C = 0$.

(e) If $1 = e_1 + \ldots + e_n \in QE(A)$ where $\{e_i\}$ is a set of orthogonal idempotents and $e_i = (m_i/n_i)f_i$ with $m_i/n_i \in Q$ and $f_i \in E(A)$ for each i then $f_1(A) \oplus \ldots \oplus f_n(A)$ is a subgroup of finite index in A.

Proof. Exercise 7.1. ///

Standard terminology for the conditions of Corollary 7.7 are:

(a) A is quasi-isomorphic to B;

(b) f is a quasi-isomorphism;

(c) A is a quasi-direct sum of $A_1, \ldots, A_n$ and each A_i is a quasi-summand of A;

(d) A is strongly indecomposable;

(e) Each f_i is a quasi-projection of A. Note that $k1_A = k_1f_1 + \ldots + k_nf_n$ for some non-zero k, $k_j \in Z$, $f_if_j = 0$ if $i \neq j$, and $m_i^2f_i^2 = n_im_if_i$. Moreover, if $nA \subseteq B \oplus C \subseteq A$ for some $0 \neq n \in Z$ then $\sigma_B = \pi_Bn$ and $\sigma_C = \pi_Cn$ are quasi-projections, where π_B and π_C are idempotent projections $B \oplus C \to B$ and $B \oplus C \to C$ respectively, since $1 = e_B + e_C$ where $e_B = (1/n)\sigma_B$ and $e_C = (1/n)\sigma_C$ are orthogonal idempotents of $QE(A)$.

Corollary 7.8: Let A be a finite rank torsion free group. Then A is strongly indecomposable iff $QE(A)$ is a local ring.

Proof. ($\Leftarrow$) Since $QE(A) = E_{QA}(A)$ is a local ring, A is indecomposable in QA.

($\Rightarrow$) $QE(A)$ is an artinian ring with no non-trivial idempotents. since idempotents split in QA. Therefore, $QE(A)$ must be a local ring (Exercise 9.1.b).

Corollary 7.9: Assume that A is finite rank torsion free.

(a) There is $0 \neq m \in Z$ with $mA \subseteq A_1 \oplus \ldots \oplus A_n \subseteq A$, where each A_i is strongly indecomposable.

(b) If $0 \neq k \in Z$ and $kA \subseteq B_1 \oplus \ldots \oplus B_\ell \subseteq A$ then each B_j is quasi-isomorphic to a direct sum of some of the A_i's.

(c) If, in addition, each B_j is strongly indecomposable then $\ell = n$ and there is a permutation σ of n^+ such that A_i is quasi-isomorphic to $B_{\sigma(i)}$ for each i.

Proof. (a) is proved by an induction on rank(A).

(b) and (c) are consequences of Theorem 7.4, Corollary 7.8, and Corollary 7.7. ///

As seen in Example 2.2 there are indecomposable groups in $\mathcal{A}$ that are not indecomposable in $Q\mathcal{A}$. Thus, although quasi-decompositions into strongly indecomposable quasi-summands are unique up to quasi-isomorphism, much of the structure of a finite rank torsion free group is lost by passing to $Q\mathcal{A}$.

Example 7.10. Let p be a prime of Z and define a category ${}_p\mathcal{A}$ by letting objects ${}_p\mathcal{A}$ = objects $\mathcal{A}$, $\mathrm{Hom}_{{}_p\mathcal{A}}(A,B) = Z_p \otimes_Z \mathrm{Hom}(A,B) \approx \mathrm{Hom}(A,B)_p$. Then ${}_p\mathcal{A}$ is an additive category such that idempotents split in ${}_p\mathcal{A}$. On the other hand, $E(A)_p = E_{{}_p\mathcal{A}}(A)$ may have only 0 and 1 as idempotents (so that A is indecomposable in ${}_p\mathcal{A}$) yet $E(A)_p$ is not a local ring.

Proof. Exercise 7.2. ///

Let p be a prime of Z and define a category $\mathcal{A}/p$ by letting the objects of $\mathcal{A}/p$ be the objects of $\mathcal{A}$ and $\mathrm{Hom}_{\mathcal{A}/p}(A,B) = \mathrm{Hom}(A,B)/p\mathrm{Hom}(A,B)$. If $f \in \mathrm{Hom}(A,B)$ then define $\overline{f} = f + p\mathrm{Hom}(A,B) \in \overline{\mathrm{Hom}(A,B)} = \mathrm{Hom}(A,B)/p\mathrm{Hom}(A,B)$. Composition is defined by letting $\overline{g}\,\overline{f} = \overline{gf}$. This category is closely related to ${}_p\mathcal{A}$ but has the additional property that $\overline{E(A)} = E_{\mathcal{A}/p}(A)$ is local iff $\overline{E(A)}$ has no idempotents other than 0 or 1. On the other hand, idempotents need not split in $\mathcal{A}/p$ (Theorem 7.11).

Let A be a torsion free group and p a prime of Z. Define $p^{\omega}A = \{x \in A \mid h_p^A(x) = \infty\}$, a pure fully invariant subgroup of A. The group A is p-reduced if $p^{\omega}A = 0$. Note that $p^{\omega}A$ is p-divisible and $A/p^{\omega}A$ is p-reduced and torsion free. Also A is a zero object in $\mathcal{A}/p$ iff $p^{\omega}A = A$ (equivalently, $pA = A$).

Theorem 7.11: (a) $\mathcal{A}/p$ is an additive category.

(b) $E_{\mathcal{A}/p}(A)$ is local iff $E_{\mathcal{A}/p}(A)$ has no idempotents other than 0 and 1_A.

(c) If A and B are isomorphic in ${}_p\mathcal{A}$ then A and B are isomorphic in $\mathcal{A}/p$. Conversely, if A and B are isomorphic in $\mathcal{A}/p$ then $A/p^{\omega}A$ and $B/p^{\omega}B$ are isomorphic in ${}_p\mathcal{A}$.

(d) If $A \simeq A_1 \oplus \ldots \oplus A_n$ in ${}_p\mathcal{A}$ then $A \simeq A_1 \oplus \ldots \oplus A_n$ in $\mathcal{A}/p$. Conversely, if $A \simeq A_1 \oplus \ldots \oplus A_n$ in $\mathcal{A}/p$, where each $A_i \neq 0$ in $\mathcal{A}/p$, then $A/p^{\omega}A \simeq A_1/p^{\omega}A_1 \oplus \ldots \oplus A_n/p^{\omega}A_n$ in ${}_p\mathcal{A}$.

(e) Idempotents need not split in $\mathcal{A}/p$.

Proof. (a) Note that $\overline{\mathrm{Hom}(A,B)}$ is an abelian group distributive with respect to composition. Let $A_1, \ldots, A_n$ be objects of $\mathcal{A}/p$ and let $A = A_1 \oplus \ldots \oplus A_n$ be the group direct sum. Then $A = A_1 \oplus \ldots \oplus A_n$ in $\mathcal{A}/p$ with injections $\{\overline{i}_1, \ldots, \overline{i}_n\}$.

(b) is a consequence of the fact that $E_{\mathcal{A}/p}(A) = E(A)/pE(A)$ is artinian.

(c) If A and B are isomorphic in ${}_p\mathcal{A}$ then there is $f \in \mathrm{Hom}(A,B)$, $g \in \mathrm{Hom}(B,A)$ and an integer n prime to p with $fg = n1_B$ and $gf = n1_A$ (Exercise 7.3). Thus, $\overline{gf} = \overline{n} \in \overline{E(A)}$ and $\overline{fg} = \overline{n} \in \overline{E(B)}$. Hence $\overline{f}$ is an isomorphism in $\mathcal{A}/p$.

Conversely, if A and B are isomorphic in $\mathcal{A}/p$ then there is $f \in \mathrm{Hom}(A,B)$, $g \in \mathrm{Hom}(B,A)$ with $gf = 1_A + ph$ and $fg = 1_B + ph'$ for some $h \in E(A)$, $h' \in E(B)$. Then $\mathrm{Kernel}(gf) \subseteq p^{\omega}A$ since if $gf(x) = x + ph(x) = 0$ then $x \in pA \cap \mathrm{Kernel}(gf) = p\mathrm{Kernel}(gf)$. Similarly, $\mathrm{Kernel}(fg) \subseteq p^{\omega}B$ so that f and g induce homomorphisms $f' : A/p^{w}A \to B/p^{w}B$ and $g' : B/p^{w}B \to A/p^{w}A$ such that $g'f'$ and $f'g'$ are monomorphisms. In fact, Image$(g'f')$ is p-pure in $A/p^{w}A$ and Image$(f'g')$ is p-pure in $B/p^{w}B$ since, for example, if $px = fg(b) = b + ph'(b)$ then $b \in pB$ and $x \in$ Image(fg). Thus, there is an integer m prime to p with $m(B/p^{w}B) \subseteq f'g'(B/p^{w}B) \subseteq f'(A/p^{w}A) \subseteq B/p^{w}B$ so that $A/p^{w}A$ and $B/p^{w}B$ are isomorphic in ${}_{p}\mathcal{A}$.

(d) is a consequence of (c) and the definition of direct sums in ${}_{p}\mathcal{A}$ and $\mathcal{A}/p$, i.e., a direct sum of $A_1, \ldots, A_n$ in ${}_{p}\mathcal{A}$ or $\mathcal{A}/p$ is isomorphic to the group direct sum with injection $\{i_1, \ldots, i_n\}$ and $\{\bar{i}_1, \ldots, \bar{i}_n\}$ respectively, where $\{i_1, \ldots, i_n\}$ are group injections.

(e) It is sufficient to find a finite rank torsion free group A such that $E(A)$ is an integral domain and a prime p such that $p^{w}A = 0$ and $E(A)/pE(A)$ has non-trivial idempotents; in which case A is strongly indecomposable hence indecomposable in $\mathcal{A}/p$ by (d).

Let $R = Z[i]$, the ring of Gaussian integers and choose an A with $5^{w}A = 0$ and $E(A) \simeq R$ (e.g., Corner's Theorem, Theorem 2.13, or the construction in Exercise 2.3). Then $5E(A) = I_1I_2$ where $I_1 = (1+2i)E(A)$ and $I_2 = (1-2i)E(A)$ are distinct maximal ideals. Write $1 = f_1 + f_2$ with $f_i \in I_i$ and note that $\bar{1} = \bar{f}_1 + \bar{f}_2$, where $\bar{f}_i^{\,2} = \bar{f}_i$ is a non-trivial idempotent for $i = 1, 2$, since $E(A)/5E(A) = I_1/5E(A) \oplus I_2/5E(A)$.

In contrast to $Q\mathcal{A}$, direct sums in $\mathcal{A}/p$ do not seem to have nice group theoretic interpretations. The following, however, is true.

<u>Corollary 7.12</u>: If $A \simeq A_1 \oplus \ldots \oplus A_n$ in $\mathcal{A}/p$ then A/pA is group isomorphic to $A_1/pA_1 \oplus \ldots \oplus A_n/pA_n$.

<u>Proof</u>. Let B be the group direct sum of $A_1, \ldots, A_n$. Then $B/pB = A_1/pA_1 \oplus \ldots \oplus A_n/pA_n$ as groups and B is isomorphic to A in $\mathcal{A}/p$, say $f \in \mathrm{Hom}(A,B)$, $g \in \mathrm{Hom}(B,A)$ with $\overline{fg} = \bar{1} \in \overline{E(B)}$ and $\overline{gf} = \bar{1} \in \overline{E(A)}$. Then f and g induce $f' : A/pA \to B/pB$ and $g' : B/pB \to A/pA$, respectively, with $f'g' = 1$ and $g'f' = 1$, i.e., A/pA and B/pB are isomorphic as groups. ///

The converse of Corollary 7.12 is false, in general (see the proof of Theorem 7.11.e).

An additive category $\mathcal{C}$ is a Krull-Schmidt category if $\mathcal{C}$ satisfies the conclusions of Theorem 7.4, e.g., $Q\mathcal{A}$ is a Krull-Schmidt category. Idempotents split in ${}_p\mathcal{A}$ yet $E_{{}_p\mathcal{A}}(A)$ may be non-local even if $E_{{}_p\mathcal{A}}(A)$ has no non-trivial idempotents. In $\mathcal{A}/p$, $E_{\mathcal{A}/p}(A)$ is local iff $E_{\mathcal{A}/p}(A)$ has no non-trivial idempotents but idempotents need not split in $\mathcal{A}/p$. The following construction gives a way out of this dilemma. The basic idea is to construct a category $(\mathcal{A}/p)^*$ which contains the "missing" summands of A in $\mathcal{A}/p$ determined by non-trivial idempotents of $E_{\mathcal{A}/p}(A)$.

Theorem 7.13: If $\mathcal{C}$ is an additive category then there is an additive category $\mathcal{C}^*$ and a functor $F : \mathcal{C} \to \mathcal{C}^*$ such that:

(a) $F : \mathrm{Hom}_{\mathcal{C}}(A,B) \to \mathrm{Hom}_{\mathcal{C}^*}(F(A), F(B))$ is an isomorphism for each A, B in $\mathcal{C}$.

(b) Idempotents split in $\mathcal{C}^*$.

(c) If X is an object of $\mathcal{C}^*$ then $E_{\mathcal{C}^*}(X) = eE_{\mathcal{C}}(A)e$ for some A in $\mathcal{C}$ and idempotent e in $E_{\mathcal{C}}(A)$.

(d) If $E_{\mathcal{C}}(A)$ is artinian for each A in $\mathcal{C}$ then $\mathcal{C}^*$ is a Krull-Schmidt category.

Proof. Let the objects of $\mathcal{C}^*$ be pairs (e, A) such that A is in $\mathcal{C}$ and e is an idempotent of $E_{\mathcal{C}}(A)$. Define $\mathrm{Hom}_{\mathcal{C}^*}((e,A), (f,B)) = \{\alpha \in \mathrm{Hom}_{\mathcal{C}}(A,B) \mid \alpha = f\alpha e\}$. Then composition is well defined and associative for if $\alpha : (e,A) \to (f,B)$ and $\beta : (f, B) \to (g, C)$ then $\beta\alpha = (g\beta f)(f\alpha e)$ so that $g(\beta\alpha)e = \beta\alpha$. Note that $e = 1_{(e,A)}$ is an identity for (e,A). Also $\mathrm{Hom}_{\mathcal{C}^*}((e,A), (f,B))$ is an abelian group distributive with respect to composition and $(0,A)$ is a zero object of $\mathcal{C}^*$.

To see that $\mathcal{C}^*$ has direct sums, let (e,A) and (f,B) be objects of $\mathcal{C}^*$ and let $C = A \oplus B$ in $\mathcal{C}$ with injections $\{i_1, i_2\}$ and projections $\{p_1, p_2\}$. Then $g = i_1ep_1 + i_2fp_2$ is an idempotent in $E_{\mathcal{C}}(C)$ so that (g,C) is an object of C^*. Moreover, $(g,C) \simeq (e,A) \oplus (f,B)$ in C^* with injections $\{gi_1e, gi_2f\}$.

If $X = (e,A)$ is in C^* then $E_{C^*}((e,A)) = eE_C(A)e$ since $\alpha : (e,A) \to (e,A)$ iff $\alpha = e\,\alpha\, e$. Also idempotents split in C^* for if $f = efe$ is an idempotent in $E_{C^*}((e,A))$ then let $q = e1_Af : (f,A) \to (e,A)$ and $p = f1_Ae : (e,A) \to (f,A)$ so that $qp = f$ and $pq = f^2 = f = 1_{(f,A)}$.

Define $F : C \to C^*$ by $F(A) = (1_A, A)$ and for $f \in \mathrm{Hom}_C(A,B)$ define $F(f) = f$. Then F is a well defined functor and $F : \mathrm{Hom}_C(A,B) \to \mathrm{Hom}_{C^*}(F(A), F(B))$ is an isomorphism.

Finally, if $E_C(A)$ is artinian for each A in C then $E_{C^*}((e,A), (e,A)) \simeq eE_C(A)e$ is artinian for each (e,A) in C^*. Thus, $X = (e,A)$ is indecomposable in C^* iff $E_{C^*}(X)$ is a local ring. Moreover, C^* is a Krull-Schmidt category by Theorem 7.4. ///

The next corollary demonstrates the somewhat surprising fact that p-reduced groups in ${}_pA$ have "cancellation" and "nth-root" properties, first proved by Lady [2].

Corollary 7.14: Suppose that A, B, C are torsion free groups of finite rank and that A and B are p-reduced.

(a) If $A \oplus C$ is isomorphic to $B \oplus C$ in ${}_pA$ then A is isomorphic to B in ${}_pA$.

(b) If A^n is isomorphic to B^n in ${}_pA$ then A is isomorphic to B in ${}_pA$.

Proof. (a) If $A \oplus C \simeq B \oplus C$ in ${}_pA$ then $A \oplus C \simeq B \oplus C$ in A/p and $(A/p)^*$ (Theorem 7.11 and 7.13). Thus, A is isomorphic to B in $(A/p)^*$, a Krull-Schmidt category by Theorem 7.13a. Moreover, $A \simeq B$ in A/p (Theorem 7.13) hence in ${}_pA$, since A and B are p-reduced (Theorem 7.11).

The proof of (b) is similar. ///

Corollary 7.15: Suppose that A, B, C are torsion free groups of finite rank and that A and B are reduced and p-local.

(a) If $A \oplus C \simeq B \oplus C$ then $A \simeq B$.

(b) If $A^n \simeq B^n$ then $A \simeq B$.

Proof. Apply Corollary 7.14 noting that A and B, being p-local, are isomorphic as groups iff they are isomorphic in ${}_p\mathcal{A}$. ///

As noted in Example 2.15, there are finite rank p-local groups with non-equivalent direct sum decompositions. The reason is, essentially, that if $A = A_1 \oplus \dots \oplus A_m = B_1 \oplus \dots \oplus B_n$ with each A_i and B_j indecomposable then A_i and B_j may not be indecomposable in $(\mathcal{A}/p)^*$. Thus, Theorem 7.13 would not necessarily imply that $m = n$ and that each A_i would be isomorphic to some B_j.

The following theorem characterizes near-isomorphism in terms of the categories ${}_p\mathcal{A}$ and $\mathcal{A}/p$. Two finite rank torsion free groups A and B are nearly isomorphic if for each $0 \neq n \in Z$ there is $m \in Z$ relatively prime to n and $f \in Hom(A,B)$, $g \in Hom(B,A)$ with $gf = m1_A$ and $fg = m1_B$. Note that isomorphism implies near isomorphism which, in turn, implies quasi-isomorphism. Moreover, A and B are nearly isomorphic iff rank(A) = rank(B) and A/d(A) and B/d(B) are nearly isomorphic, where d(A) and d(B) are the maximal divisible subgroups of A and B respectively. Thus, it is sufficient to assume that A and B are reduced.

The next theorem is essentially due to Lady [2].

Theorem 7.16: Assume that A and B are reduced torsion free groups of finite rank. The following are equivalent:

(a) $A \simeq B$ in ${}_p\mathcal{A}$ for each prime p;

(b) $A \simeq B$ in $\mathcal{A}/p$ for each prime p;

(c) For each square free integer n there are $f_n : A \to B$, $g_n : B \to A$ with $\overline{g}_n\overline{f}_n = \overline{1}_A \in E(A)/nE(A)$ and $\overline{f}_n\overline{g}_n = \overline{1}_B \in E(B)/nE(B)$

(d) For each non-zero integer n there is a monomorphism $f_n : A \to B$ and an integer m_n relatively prime to n with $m_nB \subseteq f_n(A) \subseteq B$

(e) A and B are nearly isomorphic.

Proof. (a) ⟹ (b) is Theorem 7.11.c.

(b) ⟹ (c) The groups A and B are isomorphic in A/p iff there is $f_p : A \to B$, $g_p : B \to A$ with $\overline{g}_p\overline{f}_p = \overline{1}_A \in E(A)/pE(A)$ and $\overline{f}_p\overline{g}_p = \overline{1}_B \in E(B)/pE(B)$. Thus, by induction on the number of prime divisors of the square free integer n, it is sufficient to prove that if k and n are relatively prime integers; $f_k, f_n : A \to B$; $g_k, g_n : B \to A$; $\overline{g}_k\overline{f}_k = \overline{1}_A \in E(A)/kE(A)$; $\overline{f}_k\overline{g}_k = \overline{1}_B \in E(B)/kE(B)$; $\overline{g}_n\overline{f}_n = \overline{1}_A \in E(A)/nE(A)$ and $\overline{f}_n\overline{g}_n = \overline{1}_B \in E(B)/nE(B)$ then there is $f : A \to B$, $g : B \to A$ with $\overline{gf} = \overline{1}_A \in E(A)/nkE(A)$ and $\overline{fg} = \overline{1}_B \in E(B)/nkE(B)$.

Write $1 = rk + sn$ for $r, s \in Z$ and let $f = snf_k + rkf_n$ and $g = sng_k + rkg_n$. Then $fg \equiv s^2n^2f_kg_k + r^2k^2f_ng_n \pmod{nk} \equiv snf_kg_k + rkf_ng_n \pmod{nk}$ (since $sn = snrk + s^2n^2) \equiv sn + rk \pmod{nk}$, since $f_kg_k \equiv 1 \pmod{k}$ and $f_ng_n \equiv 1 \pmod{n}$. Thus $fg \equiv 1 \pmod{nk}$. Similarly, $gf \equiv 1 \pmod{nk}$.

(c) ⟹ (d) Let n' be the product of the distinct prime divisors of n. If there is a monomorphism $f : A \to B$ and $m \in Z$ relatively prime to n' with $mB \subseteq f(A) \subseteq B$ then m is relatively prime to n. Hence it is sufficient to assume that n is square free.

Since A and B are reduced there is $0 \neq k \in Z$ with $kY \neq Y$ for any non-zero subgroup Y of A or B: let $X = A \oplus B$ and $\pi = \{p_1, p_2, \dots\}$ the set of primes. There is a descending chain of pure subgroups $p_1^wX \supseteq p_2^wp_1^wX \supseteq \dots$. Since $\operatorname{rank}(X) < \infty$ and X is reduced, $p_i^w p_{i-1}^w \dots p_1^wX = 0$ for some i. Let $k = p_ip_{i-1} \dots p_1$ so that if Y is a subgroup of X with $kY = Y$ then $Y = 0$.

It is sufficient to assume that n is square free and $nY \neq Y$ for any non-zero subgroup Y of A or B by replacing n with the product of prime divisors of nk so that if m is relatively prime to nk then m is relatively prime to n.

By (c) there is $f : A \to B$, $g : B \to A$ with $\overline{gf} = \overline{1} \in E(A)/nE(A)$ and $\overline{fg} = \overline{1} \in E(B)/nE(B)$. Write $fg = 1 + nh$ for some $h \in E(B)$. Then fg is monic since $n\text{Kernel}(fg) = \text{Kernel}(fg)$ implies $\text{Kernel}(fg) = 0$ and $\text{Image}(fg)$ is p-pure in B for each prime p dividing n. Thus there is an integer m' relatively prime to n with $m'B \subseteq fg(B) \subseteq f(A) \subseteq B$ (apply Proposition 6.1.a). Similarly, gf is a monomorphism so that f is a monomorphism.

(d) $\Rightarrow$ (e) Let $g_n = f_n^{-1} m_n$ so that $g_n f_n = m_n 1_A$ and $f_n g_n = m_n 1_B$.

(e) $\Rightarrow$ (a) is clear.

Corollary 7.17. Suppose that A, B, C are finite rank torsion free groups.

(a) If $A \oplus C$ is nearly isomorphic to $B \oplus C$ then A is nearly isomorphic to B.

(b) If A^n is nearly isomorphic to B^n then A is nearly isomorphic to B.

Proof. (a) It suffices to assume that A and B are reduced. By Theorem 7.16, $A \oplus C \simeq B \oplus C$ in $\mathcal{A}/p$, hence $(\mathcal{A}/p)^*$, for each prime p. Thus $A \simeq B$ in $(\mathcal{A}/p)^*$, hence $\mathcal{A}/p$, for each p since $(\mathcal{A}/p)^*$ is a Krull-Schmidt category (Theorem 7.13). Thus A is nearly isomorphic to B by Theorem 7.16.

The proof of (b) is similar.

Example 7.18. There are nearly isomorphic torsion free groups that are not isomorphic.

Proof. See Examples 2.10 and 2.11 and apply Corollary 7.17. ///

A finite rank torsion free group A is semi-local if $pA = A$ for all but a finite number of primes of Z.

Corollary 7.19. Suppose that A, B, C are finite rank torsion free groups and that A is semi-local.

(a) If A is nearly isomorphic to B then $A \simeq B$.

(b) If $A \oplus C \simeq B \oplus C$ then $A \simeq B$.

(c) If $A^n \simeq B^n$ then $A \simeq B$.

Proof. (a) Let $n = \pi\{p \mid pA \neq A\} \in Z$. By Theorem 7.16, there is an integer m relatively prime to n and a monomorphism $f : B \to A$ with $mA \subseteq f(B) \subseteq A$. But $mA = A$ so $f(B) = A$ and $B \simeq A$.

(b), (c) Apply Corollary 7.17 and (a). ///

The following lemma will be used in later sections.

Lemma 7.20. Suppose that A, B, C are torsion free groups of finite rank and that A is nearly isomorphic to B.

(a) If there is $f : A \to C$ and $g : C \to A$ with gf a monomorphism then $B \oplus C \simeq A \oplus A'$ for some A'.

(b) $A \oplus A' \simeq B \oplus B$ for some A' nearly isomorphic to B.

Proof. (a) Since gf is a monomorphism there is $0 \neq n \in Z$ with $nA \subseteq gf(A) \subseteq A$ (Proposition 6.1.a). Now $h = (gf)^{-1}ng : C \to A$, since $ng(C) \subseteq nA \subseteq gf(A)$ and $hf = n$. Thus, it suffices to assume $gf = n \neq 0 \in Z$. Choose an integer n' prime to n and $f' : A \to B$, $g' : B \to A$ with $g'f' = n'1_A$ and $f'g' = n'1_B$, (since A and B are nearly isomorphic). Write $1 = rn' + sn$ for $r, s \in Z$. Then $(f', f) : A \to B \oplus C$ and $(rg', sg) : B \oplus C \to A$ with $(rg', sg)(f', f) = rg'f' + sgf = rn' + sn = 1_A$ so that $B \oplus C \simeq A \oplus A'$ for some A'.

(b) follows from (a) and Corollary 7.17. ///

The following theorem is an additive category generalization of Theorem 5.1 due to Warfield [7]. Suppose that A is an object of a category $\mathcal{C}$ and that idempotents split in $\mathcal{C}$. Define $P(A)$ to be the category of summands of finite direct sums of A in $\mathcal{C}$. Then $P(A)$ is an additive category and idempotents split in $P(A)$.

Theorem 7.21: Let $\mathcal{C}$ be an additive category such that idempotents split in $\mathcal{C}$, let A be an object of $\mathcal{C}$ and let $E = E_{\mathcal{C}}(A)$. There is a category equivalence $H_A : P(A) \to P(E)$ given by $H_A(B) = \mathrm{Hom}_{\mathcal{C}}(A,B)$.

Proof. If B, C are in $P(A)$ and $f \in \mathrm{Hom}_C(B,C)$ then $H_A(f) : H_A(B) \to H_A(C)$, defined by $H_A(f)(g) = fg$ is an E-homomorphism. Moreover, H_A is a functor. If $G = B \oplus C$ in C with injections $\{i_1, i_2\}$ and projections $\{p_1, p_2\}$ then there is an E-isomorphism $H_A(G) \to H_A(B) \oplus H_A(C)$ given by $f \to (p_1 f, p_2 f)$ with inverse $(g,h) \to i_1 g + i_2 h$. Consequently, if $B \oplus C \simeq A^n$ in $P(A)$ then $H_A(B) \oplus H_A(C) \simeq H_A(A^n) \simeq E^n$ as right E-modules so that $H_A : P(A) \to P(E)$.

It is sufficient to prove that for each M in $P(E)$ there is B in $P(A)$ with $H_A(B) \simeq M$ and that for each B,C in $P(A)$ the homomorphism $H_A : \mathrm{Hom}_C(B,C) \to \mathrm{Hom}_E(H_A(B), H_A(C))$ is an isomorphism (Exercise 7.5).

Note that $H_A : E_C(A) \to E_E(H_A(A))$ is a ring isomorphism since $E_C(A) = E = H_A(A)$. Consequently, $H_A : E_C(A^m) \to E_E(H_A(A^m))$ is a ring isomorphism for each $0 < m \in Z$ being the composite of ring isomorphisms $E_C(A^m) \to \mathrm{Mat}_m(E_C(A)) \to \mathrm{Mat}_m(E_E(H_A(A))) \to E_E(H_A(A^m))$.

Suppose that B,C are in $P(A)$. Then for some $0 < n \in Z$, $B \oplus B' \simeq A^n$ and $C \oplus C' \simeq A^n$ for some B', C' in $P(A)$. As a consequence of Lemma 7.1, $\mathrm{Hom}_C(B,C) \simeq e_2\, E_C(A^n)e_1$ for some $e_i^2 = e_i \in E_C(A^n)$. Then $H_A(B) \oplus H_A(B') \simeq H_A(A^n) \simeq H_A(C) \oplus H_A(C')$ and $\mathrm{Hom}_E(H_A(B), H_A(C)) \simeq H_A(e_2)E_E(H_A(A^n))\, H_A(e_1)$. Thus, $H_A : \mathrm{Hom}_C(B,C) \to \mathrm{Hom}_E(H_A(B), H_A(C))$ is an isomorphism since $H_A : E_C(A^m) \to E_E(H_A(A^m))$ is an isomorphism.

Suppose that M is in $P(E)$. Then $M \oplus N \simeq E^n \simeq H_A(A^n)$ for some n so that $M \simeq eH_A(A^n)$ for some $e^2 = e \in E_E(H_A(A^n))$. It suffices to assume that $M = eH_A(A^n)$. Choose $f^2 = f \in E_C(A^n)$ with $H_A(f) = e$. Since idempotents split in C there is $q \in \mathrm{Hom}_C(B, A^n)$ and $p \in \mathrm{Hom}_C(A^n,B)$ with $pq = 1_B$ and $qp = f$. Then there is an isomorphism $H_A(B) \to M$ given by $g \to e(qg)$ with inverse $M \to H_A(B)$ given by $e(h) \to ph$. The latter map is well defined for if $e(h) = 0$ then $0 = fh = qph$ so that $0 = pqph = ph$. Moreover, $g \to e(qg) \to pqg = g$ and $e(h) \to ph \to e(qph) = e(fh) = e(h)$ since $e(fh) = f^2h = fh = eh$. ///

<u>Corollary 7.22.</u> Let A be a finite rank torsion free group and define $Q\mathcal{P}(A)$ to be the category of quasi-summands of finite direct sums of copies of A with $\text{Hom}_{Q\mathcal{P}(A)}(B,C) = Q\otimes_Z \text{Hom}(B,C)$ for each B,C in $Q\mathcal{P}(A)$. Then $Q\mathcal{P}(A)$ is category equivalent to $\mathcal{P}(QE(A))$.

<u>Proof.</u> Theorem 7.21 and Example 7.6. ///

EXERCISES

7.1 Prove Corollary 7.7.

7.2 Verify Example 7.10.

7.3 Prove that A and B are isomorphic in ${}_{p}\mathcal{A}$ iff there is $0 \neq n \in Z$ prime to p and $f : A \to B$, $g : B \to A$ with $fg = n1_B$ and $gf = n1_A$.

7.4 Define additive categories $\mathcal{C}_1, \mathcal{C}_2, \mathcal{C}_3, \mathcal{C}_4$ by letting the objects of each $\mathcal{C}_i$ be reduced torsion free groups of finite rank and

$$\mathrm{Hom}_{\mathcal{C}_1}(A,B) = \Pi_p \mathrm{Hom}(A,B)_p \quad ;$$

$$\mathrm{Hom}_{\mathcal{C}_2}(A,B) = \Pi_p \mathrm{Hom}(A,B)/p\mathrm{Hom}(A,B) \quad ;$$

$$\mathrm{Hom}_{\mathcal{C}_3}(A,B) = \Pi_{0<n\in Z} \mathrm{Hom}(A,B)/n\mathrm{Hom}(A,B) \quad ;$$

$$\mathrm{Hom}_{\mathcal{C}_4}(A,B) = \varprojlim_n (\mathrm{Hom}(A,B)/n\mathrm{Hom}(A,B)), \text{ the Z-adic completion of } \mathrm{Hom}(A,B) \quad .$$

(a) Prove that for each i, A and B are nearly isomorphic iff A and B are isomorphic in $\mathcal{C}_i$.

(b) Which of the $\mathcal{C}_i$'s have the property that idempotents split in $\mathcal{C}_i$?

7.5 Let $\mathcal{C}$ be an additive category such that idempotents split in $\mathcal{C}$, let A be an object of $\mathcal{C}$ and let $E = E_{\mathcal{C}}(A)$. Assume that for each M in $P(E)$ there is B in $P(A)$ with $\mathrm{Hom}_{\mathcal{C}}(A,B) = H_A(B) \simeq M$ and for each B,C in $P(A)$ the homomorphism $H_A : \mathrm{Hom}_{\mathcal{C}}(B,C) \to \mathrm{Hom}_E(H_A(B), H_A(C))$ is an isomorphism. Prove that there is a functor $S_A : P(E) \to P(A)$ such that S_AH_A is naturally equivalent to the identity functor on $P(A)$ and H_AS_A is naturally equivalent to the identity functor on $P(E)$.

7.6 (Lady) Let A and B be subgroups of a finite dimensional vector space V. The groups A and B are <u>quasi-equal</u> if $QA = QB$ and $1_A \in \mathrm{QHom}(A,B)$, $1_B \in \mathrm{QHom}(B,A)$. Prove that A and B are quasi-equal iff $QA = QB$, A_p and B_p are quasi-equal for each prime p of Z, and $A_p = B_p$ for all but a finite number of primes p of Z.

§8. Stable Range, Substitution, Cancellation, and Exchange Properties:

The basic theme of this section is to relate direct sum decompositions involving a group A to properties of E(A). Corollary 5.2 is an example of such a relationship; every B in $\mathcal{P}(A)$ is isomorphic to a direct sum of copies of A iff every finitely generated projective right E(A)-module is free.

Prior to the 1970's, the primary test for uniquenes of direct sum decompositions was the Krull-Schmidt theorem, i.e., $A = A_1 \oplus \ldots \oplus A_n$ where each $E(A_i)$ is a local ring. The primary test for cancellation, $A \oplus B \simeq A \oplus C$ implies $B \simeq C$, was to assume that A has finite exchange (Crawley-Jónsson [1]) which is the same as assuming E(A) local if A is indecomposable (Corollary 8.24). Sections 5 and 6 give some more general tests for uniqueness of direct sum decompositions while more general tests for cancellation are given in this section.

Although most of the following results are true in a more general setting (see Warfield [5]) they are stated and proved herein for torsion free groups of finite rank. The endomorphism rings of such groups have the property that every left (right) unit is a unit, a fact that will be used without further mention.

Let A be an abelian group. Then A has the <u>1-substitution property</u> if whenever $G = A \oplus B$ is an abelian group and $f \in \mathrm{Hom}(G,A)$, $g \in \mathrm{Hom}(A,G)$ with $fg = 1_A$ then there is $\phi \in \mathrm{Hom}(A,G)$ with $f\phi = 1_A$ and $G = \phi(A) \oplus B$. The group A has the <u>substitution property</u> if whenever $G = A_1 \oplus H = A_2 \oplus K$ is an abelian group with $A \simeq A_1 \simeq A_2$ then there is $C \subseteq G$ with $G = C \oplus H = C \oplus K$.

The reader is warned that "=" means equality and "≃" means isomorphism and are not to be used interchangeably in this section.

A ring R has <u>1 in the stable range</u> if whenever $f_1, f_2, g_1, g_2 \in R$ with $f_1g_1 + f_2g_2 = 1_R$ then $f_1 + f_2h$ is a unit of R for some $h \in R$ (1 in the stable range refers to the cardinal number 1).

The following theorem is due to Warfield [5] and, in part, Fuchs [5].

Theorem 8.1: Let A be an abelian group. The following are equivalent:

(a) A has 1-substitution ;

(b) 1 is in the stable range of $E(A)$;

(c) A has substitution .

Proof. (a) $\Rightarrow$ (b) Given $f_1 g_1 + f_2 g_2 = 1$ let G be an external direct sum of $A \oplus A'$ where $A' = A$, $f = (f_1, f_2) : G \to A$ and $g = (g_1, g_2) : A \to G$ ($f = (f_1, f_2)$ means that $f(a, a') = f_1(a) + f_2(a')$ for each $a \in A$, $a' \in A'$ and $g = (g_1, g_2)$ means that $g(a) = (g_1(a), g_2(a))$ where $g_1(a) \in A$ and $g_2(a) \in A'$). Then $fg = f_1 g_1 + f_2 g_2 = 1$. By (a) there is $\phi = (\phi_1, \phi_2) : A \to G$ with $f\phi = f_1\phi_1 + f_2\phi_2 = 1$ and $G = A \oplus A' = \phi(A) \oplus A'$. Note that $\phi_1 : A \to A$ is onto. Define $\theta : G \to A$ with $\theta\phi = 1_A$ and $\text{Kernel}(\theta) = A'$. Then $\theta = (h_1, 0)$, relative to $G = A \oplus A'$, with $1_A = \theta\phi = (h_1, 0)(\phi_1, \phi_2) = h_1\phi_1$ so that ϕ_1 is 1-1 hence a unit of $E(A)$. Therefore, $\phi_1^{-1} = f_1 + f_2(\phi_2\phi_1^{-1})$ is a unit of $E(A)$ with $h = \phi_2\phi_1^{-1} \in E(A)$.

(b) $\Rightarrow$ (c) It is sufficient to assume that $G = A \oplus H = A_2 \oplus K$ with $A \simeq A_2$. Let $f = (f_1, \alpha) : G \to A$ be the composition of a projection of G onto A_2 followed by an isomorphism $A_2 \to A$ with $\text{Kernel}(f) = K$ and let $g = (g_1, \beta) : A \to G$ be the inverse of $A_2 \to A \subseteq G$, where $\alpha : H \to A$; $f_1, g_1 \in E(A)$; and $\beta : A \to H$. Then $1_A = fg = f_1 g_1 + (\alpha\beta)1_A$. Since 1 is in the stable range of $E(A)$, there is a unit u of $E(A)$ and $h \in E(A)$ with $f_1 u + (\alpha\beta)h = 1_A$. Now $\theta = (u, \beta h) : A \to G$ with $1_A = f_1 u + \alpha(\beta h) = f\theta$. Thus $G = \theta(A) \oplus \text{Kernel}(f) = \theta(A) \oplus K$. Also $\phi = (u^{-1}, 0) : G \to A$, relative to $G = A \oplus H$, with $\text{Kernel}(\phi) = H$ and $1_A = \phi\theta = u^{-1}u + 0$. Therefore, $G = \theta(A) \oplus \text{Kernel}(\phi) = \theta(A) \oplus H$. Letting $C = \theta(A)$ completes the proof.

(c) =>(a) Given $G = A \oplus B$ and $f : G \to A$, $g : A \to G$ with $fg = 1_A$ then $G = g(A) \oplus \text{Kernel}(f)$ with $g(A) \simeq A$. By (c), $G = C \oplus B = C \oplus \text{Kernel}(f)$ for some $C \subseteq G$. Thus, $A \simeq C$ since $G = A \oplus B = C \oplus B$.

Let $\phi : A \to C \subseteq G$ be an isomorphism. Then $f\phi(A) = f(C) = f(C \oplus \text{Kernel}(f)) = f(G) = A$ so that $f\phi$ is a unit of $E(A)$. Thus, $\phi' = \phi(f\phi)^{-1} : A \to G$ with $f\phi' = 1_A$ and $G = C \oplus B = \phi(A) \oplus B = \phi'(A) \oplus B$, as desired. ///

Following are some examples of rings with 1 in the stable range. For a ring R define $\mathcal{J}(R)$, the Jacobson radical of R, to be the intersection of the maximal right ideals of R. Each of the following sets is equal to $\mathcal{J}(R)$: (i) the intersection of the maximal left ideals of R; (ii) $\{r \in R \mid 1-sr$ is a unit of R for all $s \in R\}$; and (iii) $\{r \in R \mid 1-rs$ is a unit of R for all $s \in R\}$ (Exercise 8.1). Consequently, $\mathcal{J}(R)$ is an ideal of R and $\mathcal{J}(R/\mathcal{J}(R)) = 0$. The ring R is <u>semi-simple</u> if $\mathcal{J}(R) = 0$.

<u>Proposition 8.2</u>: (a) (Bass [1]) If $R/\mathcal{J}(R)$ is artinian then 1 is in the stable range of R.

(b) (Goodearl [1]) Suppose that S is a subring of R with $nR \subseteq S \subseteq R$ for some $0 \neq n \in Z$ and that R/nR is finite. If 1 is in the stable range of R then 1 is in the stable range of S.

<u>Proof</u>. (a) It suffices to assume that R is semi-simple artinian for if $f_1g_1 + f_2g_2 = 1 \in R$ then $\bar{f}_1\bar{g}_1 + \bar{f}_2\bar{g}_2 = \bar{1} \in \bar{R} = R/\mathcal{J}(R)$. If $\bar{f}_1 + \bar{f}_2\bar{h} = \bar{u}$ is a unit of $\bar{R}$ for some $u, h \in R$ then $f_1 + f_2h = u + x$ for some $x \in \mathcal{J}(R)$ and $(u+x)v - 1 \in \mathcal{J}(R)$ for some $v \in R$. Therefore, $(u+x)v = 1 - (1-(u+x)v)$, hence $u + x$, is a unit of R as needed.

Assume that R is semi-simple artinian. Then $R = f_1R \oplus I$ for some right ideal $I \subseteq f_2R$ (Exercise 9.1.c). Let $\phi : R \to f_1R$ be left multiplication by f_1. Since R is semi-simple artinian, Kernel(ϕ) is a summand of R, i.e., there is $\theta : R \to \text{Kernel}(\phi)$ with $\theta|\text{Kernel}(\phi) = 1$. Hence $R = f_1R \oplus I \simeq f_1R \oplus \text{Kernel}(\phi)$. By the classical Krull-Schmidt theorem, there is an isomorphism σ : Kernel(ϕ) $\to$ I. The composite of (ϕ, θ) ; $R \to f_1R \oplus$ Kernel(ϕ) and $(1, \sigma)$: $f_1R \oplus$ Kernel(ϕ) $\to f_1R \oplus I \approx R$ is an automorphism of R with $1 \to (f_1, \theta(1)) \to (f_1, \sigma\theta(1))$, hence multiplication by $f_1 + x_1$, where $x_1 = \sigma\theta(1) \in I \subseteq f_2R$. Therefore, $f_1 + x_1$ is a unit of R. Writing $x_1 = f_2h$ shows that $f_1 + f_2h$ is a unit of R for some $h \in R$.

(b) Given $f_1, f_2, g_1, g_2 \in S$ with $f_1g_1 + f_2g_2 = 1$ let $\overline{S} = S/nR$, an artinian ring, noting that nR is an ideal of S. Then $\overline{f}_1\overline{g}_1 + \overline{f}_2\overline{g}_2 = \overline{1} \in \overline{S}$, so $\overline{f}_1 + \overline{f}_2\overline{h}$ is a unit of $\overline{S}$ for some $h \in S$ by (a).

It is sufficient to assume that $\overline{f}_1$ is a unit of $\overline{S}$, since replacing f_1 by $f_1 + f_2h$ and assuming $(f_1 + f_2h) + f_2h'$ a unit of S gives $f_1 + f_2(h+h')$ a unit of S, noting that $(f_1+f_2h)g_1 + f_2(g_2-hg_1) = 1$ satisfies the hypothesis.

Suppose that $\overline{f}_1$ is a unit of $\overline{S}$. Then $f_1\alpha + n\beta = 1$ for some $\alpha, \beta \in R$ and $f_1(\alpha + g_1n\beta) + f_2(g_2n\beta) = f_1\alpha + n\beta = 1$. It is now sufficient to assume that $g_2 \in nR$ (replace g_2 by $g_2n\beta$).

Assume that $f_1g_1 + f_2g_2 = 1$, where $\overline{f}_1$ is a unit of $\overline{S}$ and $g_2 \in nR$. Since f_1g_1 and f_2g_2 commute, $f_1g_1' + (f_2g_2)^2 = 1^2 = 1$ for some $g_1' \in R$. But 1 is in the stable range of R so that $f_1 + (f_2g_2)h$ is a unit of R for some $h \in R$, say $r(f_1+(f_2g_2)h) = (f_1+(f_2g_2)h)r = 1$. Now $f_2g_2 \in nR$ so $(r+nR)(f_1+nR) = (f_1+nR)(r+nR) = 1 + nR \in R/nR$. But $f_1 + nR$ is a unit of $\overline{S} = S/nR$ so $r + nR \in \overline{S}$ and $r \in S$. Thus $f_1 + (f_2g_2)h = f_1 + f_2(g_2h)$ is a unit of S with $g_2h \in nR \subseteq S$, as needed. ///

The following examples show that, even for subrings of Q, rings with 1 in the stable range are not easily characterized (also see Estes-Ohm [1]).

<u>Example 8.3</u>: (a) 1 is not in the stable range of Z.

(b) There is a subring R of Q such that $pR = R$ for infinitely many primes of Z yet 1 is not in the stable range of R.

(c) There is a subring R of Q such that $pR = R$ for infinitely many primes of Z, $pR \neq R$ for infinitely many primes of Z, and 1 is in the stable range of R.

Proof. (a) $3 \cdot 5 + 7(-2) = 1$ but $3 + 7h$ is not a unit of Z for any h in Z.

(b) Let $p > 3$ be a prime, $S = \{q | q \text{ is a prime } \not\equiv 1 \pmod p\}$, $R = \cap \{Z_q | q \in S\}$. Then $\{q | qR = R\} = \{q | q \equiv \pm 1 \pmod p\}$ is infinite. There are $r, s \in Z$ with $2r + ps = 1$ but $2 + ph$ is not a unit of R for any $h \in R$ since units of R are $\equiv \pm 1 \pmod p$ while $2 + ph \equiv 2 \pmod p$ and $p > 3$.

(c) Let p be a prime, $n_1, \ldots, n_k$ integers such that $\{n_i + pZ\}$ is the set of units of Z/pZ, $S' = \{q | q \text{ divides } n_i \text{ for some } i\}$, $S = \{q | q \not\equiv 1 \pmod p\} \setminus S'$, $R = \cap \{Z_q | q \in S\}$. If $f_1 g_1 + f_2 g_2 = 1$ then $f_1 + f_2 h = n_i + pr$ for some $h, r \in R$ (Proposition 8.2.a). Thus $f_1 + f_2 h$ is a unit of R. Furthermore, $\{p | pR \neq R\}$ and $\{p | pR = R\}$ are both infinite.

On the other hand:

Corollary 8.4: Assume that A is a reduced semi-local torsion free group of finite rank and let $n = \pi \{p | p \text{ is a prime with } pA \neq A\} \in Z$.

(a) $nE(A) \subseteq \mathcal{J}(E(A)) \subseteq E(A)$.

(b) 1 is in the stable range of $E(A)$.

Proof. (a) If $f \in E(A)$ then $1\text{-}nf$ is a monomorphism: Kernel$(1\text{-}nf)$ pure in A implies that Kernel$(1\text{-}nf)$ is divisible, hence 0, since A is reduced. Moreover, Image$(1\text{-}nf)$ is pure in A. Thus $1\text{-}nf$ is a unit of $E(A)$ for each $f \in E(A)$ so that $n \in \mathcal{J}(E(A))$ and $nE(A) \subseteq \mathcal{J}(E(A))$.

(b) As a consequence of (a), $E(A)/\mathcal{J}(E(A))$ is finite, hence artinian. Now apply Proposition 8.2.a. ///

There is a nice characterization of a strongly indecomposable group with 1 in the stable range of its endomorphism ring.

Corollary 8.5: Suppose that A is a finite rank torsion free group.

(a) If 1 is in the stable range of $E(A)$ and if $n \in Z$ then each unit of $E(A)/nE(A)$ lifts to a unit of $E(A)$.

(b) If A is strongly indecomposable and if each unit of $E(A)/nE(A)$ lifts to a unit of $E(A)$ for each $n \in Z$ then 1 is in the stable range of $E(A)$.

Proof. (a) If $\bar{f}$ is a unit of $\bar{E} = E(A)/nE(A)$ then $fg + nh = 1$ for some $g,h \in E(A)$. Since 1 is in the stable range of $E(A)$, $f + nh_1 = u$ is a unit of $E(A)$ for some $h_1 \in E(A)$ and $\bar{f} = \bar{u}$.

(b) Recall that $QE(A)$ is a local ring if A is strongly indecomposable (Corollary 7.8). Thus, $\mathcal{J}(QE(A))$ is the unique maximal ideal of non-units of $QE(A)$ and is nilpotent since $QE(A)$ is artinian. Hence if $f \in E(A)$ then either f is in $\mathcal{J}(QE(A))$, hence nilpotent, or else $f \notin \mathcal{J}(QE(A))$ is a unit of $QE(A)$.

Suppose that $f_1g_1 + f_2g_2 = 1$. If f_2g_2 is nilpotent then $f_1g_1 = 1 - f_2g_2$ is a unit of $E(A)$ so that $f_1 = f_1 + f_2 \cdot 0$ is a unit of $E(A)$.

If f_2g_2 is not nilpotent then f_2g_2, hence f_2 is a unit in $QE(A)$, say $f_2f_2' = n \in Z$ is non-zero for some $f_2' \in E(A)$. Now $\bar{f}_1\bar{g}_1 + \bar{f}_2\bar{g}_2 = \bar{1} \in \bar{E} = E(A)/nE(A)$ so $\bar{f}_1 + \bar{f}_2\bar{h}$ is a unit of $\bar{E}$ for some $h \in E$ (Proposition 8.2). By hypothesis, $f_1 + f_2h = u + nh'$ for some unit u of $E(A)$ and $h' \in E(A)$. Thus, $u = f_1 + f_2h - nh' = f_1 + f_2(h-f_2'h')$ is a unit of $E(A)$. ///

Corollary 8.6. Suppose that A is finite rank torsion free strongly indecomposable and that $E(A) \to E(A/nA)$ is onto for each $n \in Z$ (e.g. rank$(A) = 1$). Then 1 is in the stable range of $E(A)$ iff for each $n \in Z$, every automorphism of A/nA lifts to an automorphism of A.

Corollary 8.7. Assume that A and B are finite rank torsion free abelian groups.

(a) Suppose that $A = B \oplus C$. Then 1 is in the stable range of $E(A)$ iff 1 is in the stable range of both $E(B)$ and $E(C)$.

(b) Suppose that B is isomorphic to a fully invariant subgroup of finite index in A. If 1 is in the stable range of $E(B)$ then 1 is in the stable range of $E(A)$.

(c) Suppose that A and B are nearly isomorphic. If 1 is in the stable range of $E(B)$ then 1 is in the stable range of $E(A)$.

Proof. (a) In view of Theorem 8.1, it is sufficient to prove that A has the substitution property iff both B and C have the substitution property.

Assume that B and C have the substitution property and let $G = A_1 \oplus H = A_2 \oplus K$ with $A \simeq A_1 \simeq A_2$. Write $A_i = B_i \oplus C_i$ with $B_i \simeq B$, $C_i \simeq C$ so that $G = B_1 \oplus C_1 \oplus H = B_2 \oplus C_2 \oplus K$. Since B has substitution, $G = D_1 \oplus C_1 \oplus H = D_1 \oplus C_2 \oplus K$ for some D_1. Thus $G = (D_2 \oplus D_1) \oplus H = (D_2 \oplus D_1) \oplus K$ since $C \simeq C_1 \simeq C_2$ has substitution.

Conversely, assume that A has substitution and let $G = B_1 \oplus H = B_2 \oplus K$ with $B \simeq B_1 \simeq B_2$. Define $G' = C \oplus B_1 \oplus H = C \oplus B_2 \oplus K$ so that $A \simeq C \oplus B_1 \simeq C \oplus B_2$. Then $G' = D \oplus H = D \oplus K$ for some D and $G = G \cap G' = (D \cap G) \oplus H = (D \cap G) \oplus K$, since $H, K \subseteq G \subseteq G'$. Therefore B, and similarly C, has substitution.

(b) Assume that $nA \subseteq B \subseteq A$ for some $0 \neq n \in Z$, where B is a fully invariant subgroup of A. There is a ring monomorphism $\phi : E(A) \to E(B)$, defined by $\phi(f) = f|_B$, and $nE(B) \subseteq \text{Image}(\phi) \subseteq E(B)$. Now apply Proposition 8.2.b.

(c) Since A is nearly isomorphic to B, $A \oplus A' \simeq B \oplus B$ for some A' (Lemma 7.20). Thus, (a) applies.

An abelian group A has the cancellation property if whenever $A \oplus B \simeq A \oplus C$ then $B \simeq C$.

Corollary 8.8. (a) If A has the substitution property then A has the cancellation property.

(b) Z has the cancellation property but 1 is not in the stable range of $E(Z) \simeq Z$, hence Z does not have the substitution property.

Proof. (a) Assume that $G = A \oplus B \simeq A \oplus C$. Then $G = A \oplus B = A_1 \oplus C_1$ for some $A_1 \simeq A$; $C_1 \simeq C$. Then $G = D \oplus B = D \oplus C$ for some D so that $B \simeq C$.

(b) Example 8.3.a shows that 1 is not in the stable range of Z. To show that Z has cancellation, assume $G = A_1 \oplus B = A_2 \oplus C$ with $A_1 \simeq A_2 \simeq Z$. It suffices to assume $B \not\subseteq C$ and $C \not\subseteq B$. Then there are exact sequences $0 \to B \cap C \to B \xrightarrow{\pi_2} A_2$ and $0 \to B \cap C \to C \xrightarrow{\pi_1} A_1$ induced by projections of G onto A_2 and A_1, respectively. Since $A_2 \simeq A_1 \simeq Z$, $\pi_2(B) \simeq Z \simeq \pi_1(C)$ and $B \simeq B \cap C \oplus Z \simeq C$.

Corollary 8.9. Let A, B, C be reduced finite rank torsion free groups with $A \oplus B \simeq A \oplus C$.

(a) If A is semi-local then $B \simeq C$.

(b) If B is semi-local then $B \simeq C$.

Proof. (a) Apply Corollary 8.4, Theorem 8.1, and Corollary 8.8.a.

(b) Apply Corollary 7.19.

The following results are devoted to the examination of a converse to Corollary 8.8.a, due in part to Fuchs [5] and Fuchs-Loonstra [2].

Lemma 8.10 (Fuchs [5]). Let A be a finite rank torsion free group and let $0 \neq n \in Z$ with $nA \neq A$. Assume there is a finite rank torsion free group G with $E(G) \simeq Z$, $\mathrm{Hom}(A,G) = \mathrm{Hom}(G,A) = 0$, and an epimorphism $\beta : G \to A/nA$. If A has cancellation then every unit of $E(A)/nE(A)$ lifts to a unit of $E(A)$.

Proof. Let $\alpha : A \to A/nA$ be the canonical homomorphism.

Let $\overline{f}$ be a unit of $\overline{E(A)} = E(A)/nE(A)$ and define $\alpha' = \alpha f : A \to A/nA$. Then α' is an epimorphism since there is $\overline{g} \in E(A)/nE(A)$ with $\overline{fg} = \overline{1} \in \overline{E(A)}$. Define $M = \{(a,x) \in A \oplus G \mid \alpha(a) = \beta(x)\}$ and $M' = \{(a,x) \in A \oplus G \mid \alpha'(a) = \beta(x)\}$, pull-backs of $\{\beta, \alpha\}$ and $\{\beta, \alpha'\}$, respectively. Then $\alpha\lambda = \beta\gamma$ where $\lambda : M \to A$ is defined by $\lambda(a,x) = a$ and $\gamma : M \to G$ is defined by $\gamma(a,x) = x$.

Moreover, $\text{Kernel}(\lambda) \simeq \text{Kernel}(\beta)$ and $\text{Kernel}(\gamma) \simeq nA \simeq A$. Similarly, there is $\lambda' : M' \to A$ and $\gamma' : M' \to G$ with $\alpha'\lambda' = \beta\gamma'$, $\text{Kernel}(\lambda') \simeq \text{Kernel}(\beta)$ and $\text{Kernel}(\gamma') \simeq A$.

The universal property of pull-backs implies the existence of $\sigma : M \to M'$, $\sigma' : M' \to M$ and commutative diagrams:

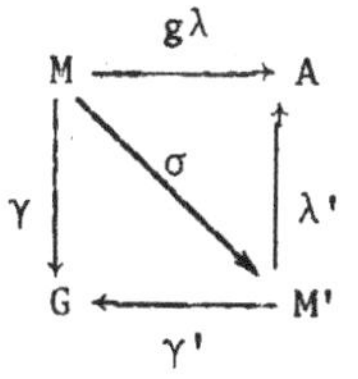

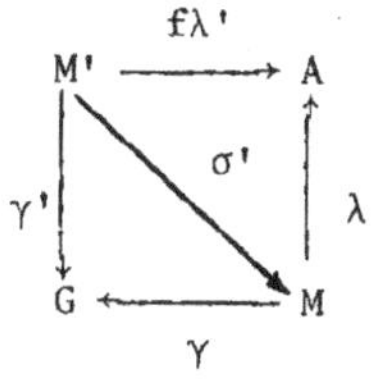

since $\alpha' g\lambda = \alpha f g\lambda = \alpha\lambda = \beta\gamma$ and $\alpha f\lambda' = \alpha'\lambda' = \beta\gamma'$. Let K be a pull-back of $\gamma' : M' \to G$ and $\gamma : M \to G$, say $\delta : K \to M$ and $\delta' : K \to M'$ with $\gamma'\delta' = \gamma\delta$. Then $K \simeq M \oplus \text{Kernel}(\gamma')$ since $K = \{(m, m') \in M \oplus M' \mid \gamma(m) = \gamma'(m')\}$; $\delta : K \to M \to 0$, given by $\delta(m, m') = m$ is onto, since α' hence γ' is onto; $\text{Kernel}(\delta) \simeq \text{Kernel}(\gamma')$; there is $h : M \to K$ given by $h(m) = (m, \sigma(m)) \in K$, since $\gamma(m) = \gamma'\sigma(m)$; and $\delta h(m) = \delta(m, \sigma(m)) = m$ for each $m \in M$.

Similarly, $K \simeq M' \oplus \text{Kernel}(\gamma)$ so that $M' \oplus A \simeq M \oplus A$, since $A \simeq \text{Kernel}(\gamma') \simeq \text{Kernel}(\gamma)$. Since A has cancellation there is an isomorphism $\mu : M' \to M$.

Since $\text{Hom}(A,G) = \text{Hom}(G,A) = 0$ there are commutative diagrams:

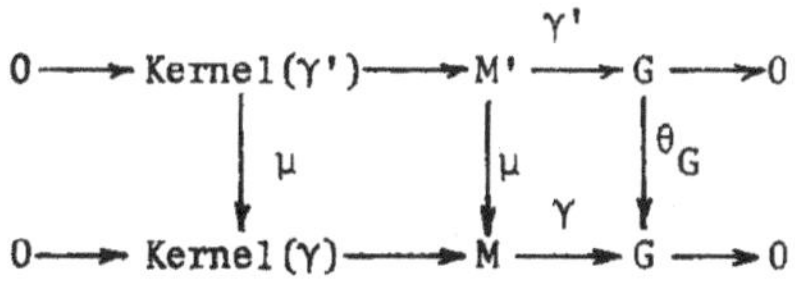

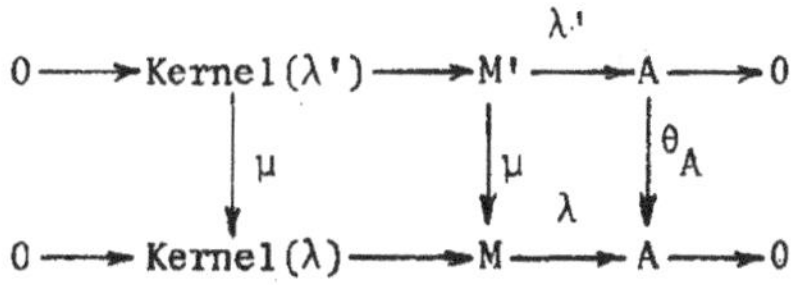

where θ_G is an automorphism of G and θ_A is an automorphism of A. If $a \in A$ then $\alpha'(a) = \beta(x)$ for some $x \in G$, since β is onto, whence $(a,x) \in M'$. Thus $\mu(a,x) = (\theta_A(a), \theta_G(x))$ since $\gamma\mu(a,x) = \theta_G\gamma'(a,x) = \theta_G(x)$ and $\gamma\mu(a,x) = \theta_A\lambda'(a,x) = \theta_A(a)$. But $\mu(a,x) \in M$ so that $\alpha\theta_A(a) = \beta\theta_G(x)$. Since $E(G) \simeq Z$, $\theta_G = \pm 1$. Consequently, $\alpha f(a) = \alpha'(a) = \beta(x) = \pm\alpha\theta_A(a)$ for each $a \in A$ so that $\overline{f}$ lifts to $\theta = \pm\theta_A$, an automorphism of A with $\overline{f} = \overline{\theta}$.

Lemma 8.11 (Fuchs-Loonstra [2]). Let A be a rank-1 torsion free group $0 \neq n \in Z$ with $nA \neq A$. If $A \not\cong Z$ and if A is not semi-local then there is a rank-1 torsion free group G with $E(G) \simeq Z$, $\mathrm{Hom}(A,G) = \mathrm{Hom}(G,A) = 0$, and an epimorphism $\beta : G \to A/nA$.

Proof. Let $\mathrm{type}(A) = [(k_p)]$. Since A is not semi-local, $S = \{p \mid 0 < k_p < \infty\}$ is infinite. Write $S = S_o \cup S_1$ as the union of disjoint infinite subsets. Define a rank-1 group G by $\mathrm{type}(G) = [(\ell_p)]$ where $\ell_p = k_p + 1$ if $p \in S_o$, $\ell_p = k_p - 1$ if $p \in S_1$ and $\ell_p = 0$ if $p \in S_1$. Then $E(G) \simeq Z$ (Theorem 1.5) and $\mathrm{Hom}(A,G) = \mathrm{Hom}(G,A) = 0$ since $\mathrm{type}(G)$ and $\mathrm{type}(A)$ are incomparable. Moreover, there is an epimorphism $G \to G/nG \to A/nA$ since $pA = A$ iff $k_p = \infty$.

Theorem 8.12 (Fuchs-Loonstra [2]). Assume that A is a rank-1 torsion free group. Then A has cancellation iff $A \simeq Z$ or else 1 is in the stable range of $E(A)$.

Proof. ($\Leftarrow$) Corollary 8.8 and Theorem 8.1.

($\Rightarrow$) Assume that $A \neq Z$. If A is semi-local then 1 is in the stable range of $E(A)$ by Corollary 8.4. If A is not semi-local then apply Lemma 8.11, 8.10, and Corollary 8.5.b.

A ring R has 2 in the stable range of R if whenever $f_1g_1 + f_2g_2 + f_3g_3 = 1 \in R$ then $1 = (f_1+f_3h_1)k_1 + (f_2+f_3h_2)k_2$ for some $h_1, h_2, k_1, k_2 \in R$. If 1 is in the stable range of R then 2 is in the stable range of R: for if $f_1g_1 + (f_2g_2+f_3g_3) = 1$ then $f_1 + (f_2g_2+f_3g_3)h_1 = u_1$ is a unit of R for some $h_1 \in R$. Thus, $(f_1 + f_3(g_3h_1))u_1^{-1} + (f_2 + f_3 \cdot 0)(g_2h_1u_1^{-1}) = 1$ as needed.

An abelian group A has 2-substitution if whenever $G = A \oplus A \oplus B$ is an abelian group and $f : G \to A$, $g : A \to G$ with $fg = 1_A$ then there is $\phi : A \to G$ such that $f\phi = 1_A$ and $G = \phi(A) \oplus C \oplus B$ for some $C \subseteq A \oplus A$.

Theorem 8.13 (Warfield [5]). Suppose that A is an abelian group. The following are equivalent:

(a) 2 is in the stable range of E(A);

(b) A has 2-substitution;

(c) If $f_1g_1 + f_2g_2 + f_3g_3 = 1 \in E(A)$ then there is $a_1, a_2, a_3 \in E(A)$ with $f_1a_1 + f_2a_2 + f_3a_3 = 1$ and $b_1a_1 + b_2a_2 = 1$ for some $b_1, b_2 \in E(A)$.

Proof. (a)=>(b) Given $G = A \oplus A \oplus B$, $f = (f_1, f_2, \alpha) : G \to A$ and $g = (g_1, g_2, \beta) : A \to G$ with $1_A = fg = f_1g_1 + f_2g_2 + (\alpha\beta)1_A$ for some $\alpha : B \to A$ and $\beta : A \to B$ then $(f_1 + \alpha\beta h_1)k_1 + (f_2 + \alpha\beta h_2)k_2 = 1_A$ for some $h_i, k_i \in E(A)$, by (a).

Define $\phi = (k_1, k_2, \beta(h_1k_1+h_2k_2)) : A \to G$ so that $f\phi = f_1k_1 + f_2k_2 + \alpha\beta(h_1k_1+h_2k_2) = 1_A$. Moreover, $\theta = (f_1+\alpha\beta h_1, f_2+\alpha\beta h_2, 0) : G \to A$ with $\theta\phi = (f_1+\alpha\beta h_1)k_1 + (f_2+\alpha\beta h_2)k_2 = 1_A$. Thus, $G = \phi(A) \oplus \text{Kernel}(\theta)$. But $B \subseteq \text{Kernel}(\theta)$ implies that $\text{Kernel}(\theta) = B \oplus C$, where $C = (A \oplus A) \cap \text{Kernel}(\theta) \subseteq A \oplus A$. Thus, $G = \phi(A) \oplus B \oplus C$ and $f\phi = 1_A$.

(b)=>(c) Given $f_1g_1 + f_2g_2 + f_3g_3 = 1 \in E(A)$ let $G = A \oplus A \oplus A$, $f = (f_1, f_2, f_3) : G \to A$ and $g = (g_1, g_2, g_3) : A \to G$ so that $fg = f_1g_1 + f_2g_2 + f_3g_3 = 1_A$. By (b), there is $\phi = (h_1, h_2, h_3) : A \to G$ with $1_A = f\phi = f_1h_1 + f_2h_2 + f_3h_3$ and $G = \phi(A) \oplus C \oplus A$ for some $C \subseteq A \oplus A \oplus 0$. Choose $\theta : G \to A$ with $\theta\phi = 1_A$ and $\text{Kernel}(\theta) = C \oplus A$. Then $\theta = (k_1, k_2, 0) : G \to A$ with $\theta\phi = k_1h_1 + k_2h_2 + 0\,h_3 = 1$, as needed.

(c)=>(a) Given $f_1g_1 + f_2g_2 + f_3g_3 = 1_A$ then there are $a_i, b_i \in E(A)$ with $f_1a_1 + f_2a_2 + f_3a_3 = 1 = b_1a_1 + b_2a_2$, by (c). Thus, $(f_1+f_3(a_3b_1))a_1 + (f_2+f_3(a_3b_2))a_2 = f_1a_1 + f_2a_2 + f_3a_3(b_1a_1+b_2a_2) = f_1a_1 + f_2a_2 + f_3a_3 = 1_A$. ///

There are definitions of n in the stable range of E(A) and n-substitution which are equivalent for $n \geq 1$ (Warfield [5]). The following theorem shows that 2 is in the stable range of E(A) for any finite rank torsion free group A so that it is not necessary, in these notes, to consider n in the stable range and n-substitution for $n \geq 3$.

Lemma 8.14 (Warfield [5]). Assume that A is torsion free of finite rank and that $f_1, f_2, g_1, g_2 \in E(A)$. If $f_1g_1 + f_2g_2$ is a monomorphism then there is $h \in E(A)$ such that $f_1 + f_2h$ is a monomorphism.

Proof. If $f \in E(A)$ then f is a monomorphism iff f is a unit of $QE(A)$ (Proposition 6.1.a). If f is a unit of $QE(A)$ then $f + \mathcal{J}(QE(A))$ is not a zero divisor of $\overline{QE(A)} = QE(A)/\mathcal{J}(QE(A))$. Conversely, if $f \in E(A)$ and $\overline{f} = f + \mathcal{J}(QE(A))$ is not a left zero divisor in $\overline{QE(A)}$ then $\overline{f}$ must be a unit of $\overline{QE(A)}$, since $\overline{QE(A)}$ is semi-simple artinian, so that f is a unit of $QE(A)$.

Thus, it is sufficient to assume that $\mathcal{J}(QE(A)) = 0$, $f_1g_1 + f_2g_2$ is a unit of $QE(A)$ and to prove that $f_1 + f_2h$ is not a left zero divisor of $QE(A)$ for some $h \in E(A)$. As in Proposition 8.2.a, $QE(A) = f_1QE(A) \oplus I$ for some right ideal $I \subseteq f_2QE(A)$ and $f_1 + f_2h' = 1$ for some $h' \in QE(A)$ with $f_2h' \in I$. Write $h' = (1/m)h$ for $0 \neq m \in Z$, $h \in E(A)$. Then $m = mf_1 + f_2h \neq 0$. Therefore, $f_1 + f_2h$ is not a left zero divisor of $QE(A)$ for if $(f_1+f_2h)r = 0$ for some $r \in QE(A)$ then $0 = f_1r + f_2hr \in f_1QE(A) \oplus I \subseteq QE(A)$ so that $0 = f_1r = f_2hr$, $mr = (mf_1+f_2h)r = 0$ and $r = 0$ since $E(A)$ is torsion free. ///

Theorem 8.15 (Warfield [5]). Suppose that A is a finite rank torsion free group.

(a) 2 is in the stable range of $E(A)$.

(b) If $A \oplus A \oplus X \simeq A \oplus Y$ then $Y \simeq A' \oplus X$, where $A \oplus A' \simeq A \oplus A$.

Proof. (a) Assume that $f_1g_1 + f_2g_2 + f_3g_3 = 1_A$, where each $f_i, g_i \in E(A)$. Then $f_1 + (f_2g_2+f_3g_3)h$ is a monomorphism for some $h \in E(A)$ by Lemma 8.14. Let $f_1' = (f_2g_2+f_3g_3)h = f_2\alpha_2 + f_3\alpha_3$, where $\alpha_2 = g_2h$, and $\alpha_3 = g_3h$. Then $f_1 + f_1'$ is monic and $E(A)/(f_1+f_1')E(A)$ is finite (Proposition 6.1.a).

It is sufficient to assume that $nE(A) \subseteq f_1E(A) \subseteq E(A)$ for some $0 \neq n \in Z$, since replacing f_1 by $f_1 + f_1'$ gives $(f_1+f_1')g_1 + f_2(g_2-\alpha_2g_1) + f_3(g_3-\alpha_3g_1) = f_1g_1 + f_2g_2 + f_3g_3 = 1_A$. Thus, if $((f_1+f_1') + f_3h_1)k_1 + (f_2+f_3h_2)k_2 = 1$ for some $h_i, k_i \in E(A)$ then $f_1k_1 + f_2(\alpha_2k_1+k_2) + f_3(\alpha_3k_1+h_1k_1+h_2k_2) = 1$ with $(f_1+f_2\alpha_2+f_3\alpha_3+f_3h_1-f_2\alpha_2-f_3h_2\alpha_2)k_1 + (f_2+f_3h_2)(\alpha_2k_1+k_2) = 1$ so that an application of Theorem 8.13 (c)$\Longrightarrow$(a) completes the proof.

Let $J_n/nE(A) = \mathcal{J}(E(A)/nE(A))$. Then $E(A)/J_n$ is semi-simple artinian so that 1 is in the stable range of $E(A)/nE(A) = \overline{E}$ (Proposition 8.2.a). But $\overline{f_2g_2} + \overline{f_1g_1} + \overline{f_3g_3} = \overline{1} \in \overline{E}$ so $\overline{f_2} + \overline{(f_1g_1+f_3g_3)h} = \overline{u}$, a unit of $\overline{E}$ for some $h \in E(A)$. Thus, $(f_2+f_1E(A)) + (f_3g_3h+f_1E(A)) = u + f_1E(A) \in E(A)/f_1E(A)$, since $nE(A) \subseteq f_1E(A)$. Choosing v with $uv - 1 \in nE(A) \subseteq f_1E(A)$ gives $(f_2+f_3g_3h)v + f_1E(A) = 1 + f_1E(A) \in E(A)/f_1E(A)$. Therefore, $1 = f_1k_1 + (f_2+f_3g_3h)v = (f_1+f_3\cdot 0)k_1 + (f_2+f_3(g_3h))v$ for some $k_1 \in E(A)$ whence 2 is in the stable range of $E(A)$.

(b) Let $G = A_1 \oplus A_2 \oplus X = A \oplus Y$, where $A_1 \simeq A_2 \simeq A$ and let $f : G \to A$ be a projection with Kernel$(f) = Y$. By (a) and Theorem 8.13, A has the 2-substitution property, i.e., $G = A_1 \oplus A_2 \oplus X = \phi(A) \oplus \text{Kernel}(f) = \phi(A) \oplus Y$ $\phi(A) \oplus A' \oplus X$ for some $A' \subseteq A_1 \oplus A_2$. Thus, $Y \simeq A' \oplus X$ and $A \oplus A \simeq A_1 \oplus A_2 \simeq \phi(A) \oplus A' \simeq A \oplus A'$. ///

An abelian group A has <u>self-cancellation</u> if $A \oplus A' \simeq A \oplus A$ implies that $A' \simeq A$. As a consequence of the proof of Theorem 8.1, 1 is in the stable range of $E(A)$ iff whenever $G = A_1 \oplus A_2 = A \oplus A'$ with $A_1 \simeq A_2 \simeq A$ then $G = C \oplus A' = C \oplus A_2$ for some C. Thus, if A is finite rank torsion free, 1 in the stable range of $E(A)$ implies self-cancellation and 2 is always in the stable range of $E(A)$.

Theorem 8.15, together with some self-cancellation hypotheses, gives cancellation in certain cases.

<u>Corollary 8.16</u>. Suppose that A, B, C are torsion free groups of finite rank and that $A \oplus B \simeq A \oplus C$.

(a) If A is isomorphic to a summand of B and if A has self-cancellation then $B \simeq C$.

(b) If $B^n = A_1 \oplus X$ for some $0 < n \in Z$, where A_1 is nearly isomorphic to A, and if B has self-cancellation then $B \simeq C$.

Proof. (a) If $B \simeq A \oplus X$ then, by Theorem 8.15.b, $C \simeq A' \oplus X$ for some A' with $A \oplus A' \simeq A \oplus A$. Since A has self-cancellation, $A \simeq A'$. Thus $C \simeq A \oplus X \simeq B$.

(b) If $B^n = A_1 \oplus X$ then $B^{2n} = A_1 \oplus A_1 \oplus X \oplus X$. But $A \oplus A_2 \simeq A_1 \oplus A_1$ for some A_2, since A is nearly isomorphic to A_1 (Lemma 7.20). Thus it suffices to assume $B^n \simeq A \oplus X$, replacing $2n$ by n. Now $B^{n+1} \simeq X \oplus A \oplus B \simeq X \oplus A \oplus C = B^n \oplus C$. If $n > 1$ then $B^{n-1} \oplus C \simeq B^{n-1} \oplus B'$ for some B' with $B \oplus B' \simeq B \oplus B$ (Theorem 8.15.b). Thus, $B^{n-1} \oplus C \simeq B^{n-1} \oplus B' \simeq B^n$, thereby reducing to the case $n = 1$, i.e., $B \oplus B \simeq B \oplus C$. since B has self-cancellation, $B \simeq C$. ///

Finite rank torsion free groups with self-cancellation are abundant, in contrast to groups with 1 in the stable range.

Corollary 8.17. Let A be a finite rank torsion free group.

(a) A has self-cancellation iff whenever $E(A) \oplus M \simeq E(A) \oplus E(A)$ as right $E(A)$-modules then $M \simeq E(A)$. In particular if finitely generated projective right $E(A)$-modules are free then A has self-cancellation.

(b) If $A \simeq B^2$ then A has self-cancellation.

(c) If $A = B \oplus C$ and B,C have self-cancellation then A has self-cancellation.

(d) If $E(A)$ is commutative then A has self-cancellation.

Proof. (a) is a consequence of the duality given in Theorem 5.1.

(b) Suppose $A \oplus A' \simeq A \oplus A$ so that $B^2 \oplus A' \simeq B^4$. Then $B \oplus A' \simeq B' \oplus B^2$, where $B \oplus B' \simeq B \oplus B$ (Theorem 8.15.b). Thus, $B \oplus A' \simeq B^3$. Again by Theorem 8.15.b, $A' \simeq B \oplus B'$, where $B \oplus B' \simeq B \oplus B$, i.e., $A' \simeq B \oplus B \simeq A$.

(c) Suppose that $A \oplus A' \simeq A \oplus A$. Then $B \oplus C \oplus A' \simeq B \oplus C \oplus B \oplus C$. By Theorem 8.15.b, $C \oplus A' \simeq B' \oplus C \oplus C$, where $B \oplus B' \simeq B \oplus B$. Since B has self-cancellation, $B' \simeq B$ so that $C \oplus A' \simeq B \oplus C \oplus C$. Similarly, using C instead of B, $A' \simeq B \oplus C'$ where $C \oplus C' \simeq C \oplus C$ and $C' \simeq C$. Thus, $A' \simeq B \oplus C \simeq A$.

(d) Let $\theta = \begin{pmatrix} f_1 & f_2 \\ g_1 & g_2 \end{pmatrix} : A \oplus A \to A \oplus A'$ be an isomorphism with $f_i : A \to A$, $g_i : A \to A'$ having an inverse $\begin{pmatrix} h_1 & k_1 \\ h_2 & k_2 \end{pmatrix} : A \oplus A' \to A \oplus A$, with $h_i : A \to A$, $k_i : A' \to A$. Since A' is nearly isomorphic to A (Corollary 7.17) it suffices to assume that $f_i, g_i, h_i, k_i \in QE(A)$. Thus, $(\det \theta)(A) = (g_2f_1-g_1f_2)(A) \subseteq A'$ and $(\det \theta^{-1})A' = (h_1k_2-h_2k_1)(A') \subseteq A$. It follows that $h_1k_2 - h_2k_1 : A' \to A$ is an isomorphism. ///

A finite dimensional simple Q-algebra K is a <u>totally definite quaternion algebra</u> if K is a division algebra with center F such that $\dim_F K = 4$ and if ν is an archimedean valuation of F, then the ν-completion, F_ν, of F is isomorphic to the reals and $K_\nu = F_\nu \otimes_F K$ is the ring of Hamiltonian quaternions.

<u>Theorem 8.18</u>: Suppose that A is a finite rank torsion free group and let $QE(A)/\mathcal{J}(QE(A) = K_1 x \ldots x K_n$ be a product of simple Q-algebras. If no K_i is a totally definite quaternion algebra then A has self-cancellation.

<u>Proof</u>. Arnold [7]. ///

As an application of Theorem 8.18:

<u>Corollary 8.19</u>: If A is an almost completely decomposable torsion free group of finite rank then A has self-cancellation.

<u>Proof</u>. By Theorem 9.10 $QE(A)/\mathcal{J}(QE(A))$ is a product of matrix rings over Q. Furthermore, matrix rings over Q are simple Q-algebras but not totally definite quaternion algebras. Now apply Theorem 8.18. ///

<u>Example 8.20</u>. There is a finite rank torsion free group that fails to have self-cancellation.

<u>Proof</u>. By Swan [1] there is a ring R, a subring of the group ring $Q(G)$ where G is the generalized quaternion group of order 32, and a non-principal ideal I with $R \oplus I \simeq R \oplus R$. By Corner's Theorem (Theorem 2.13), $R \simeq E(A)$ for some finite rank torsion free A.

Now A fails to have self-cancellation by Corollary 8.17.a. ///

As a consequence of Example 8.20 and Corollary 8.17.b, summands of groups with self-cancellation need not have self-cancellation.

An abelian group A has the n-exchange property if whenever $G = A \oplus B = C_1 \oplus \dots \oplus C_n$ is an abelian group then $G = A \oplus C_1' \oplus \dots \oplus C_n'$ for some $C_i' \subseteq C_i$, $1 \leq i \leq n$ and the finite exchange property if A has the n-exchange property for all $1 < n \in Z$. The next lemma is due to Crawley-Jónsson [1].

Lemma 8.21. Let A be an abelian group.

(a) Assume that $A = B \oplus C$. Then A has n-exchange iff B and C have n-exchange.

(b) If A has 2-exchange then A has finite exchange.

Proof. (<=) Suppose that $G = A \oplus D = C_1 \oplus \dots \oplus C_n = B \oplus C \oplus D$. Since C has n-exchange, $G = A \oplus D = B \oplus C \oplus D = C \oplus C_1' \oplus \dots \oplus C_n'$ for some $C_i' \subseteq C_i$. Thus, $B \oplus D \simeq C_1' \oplus \dots \oplus C_n' = B' \oplus D'$ with $B \simeq B'$, $D \simeq D'$. Now B, hence B', has n-exchange so $B' \oplus D' = B' \oplus C_1'' \oplus \dots \oplus C_n''$ for some $C_i'' \subseteq C_i' \subseteq C_i$. Hence, $G = A \oplus D = C \oplus C_1' \oplus \dots \oplus C_n' = C \oplus B' \oplus C_1'' \oplus \dots \oplus C_n''$, where $C_i'' \subseteq C_i$. But $p_B i_B : B \to B'$ is an isomorphism, where $i_B : B \to G$ is an injection and $p_B : G \to B'$ is a projection. By Lemma 7.2.d, B' can be replaced by B, i.e. $G = A \oplus D = C \oplus B \oplus C_1'' \oplus \dots \oplus C_n'' = A \oplus C_1'' \oplus \dots \oplus C_n''$.

(=>) To see, for example, that B has n-exchange, let $G = B \oplus D = C_1 \oplus \dots \oplus C_n$ and $G' = A \oplus D = C \oplus B \oplus D = (C \oplus C_1) \oplus \dots \oplus C_n = C \oplus G$. Since A has n-exchange, $G' = A \oplus D = C \oplus B \oplus D = A \oplus E_1' \oplus C_2' \oplus \dots \oplus C_n' = C \oplus B \oplus E_1' \oplus C_2' \oplus \dots \oplus C_n'$ for some $E_1' \subseteq C \oplus C_1$ and $C_i' \subseteq C_i$, $2 \leq i \leq n$. Thus, $G = G \cap G' = B \oplus ((C \oplus E_1') \cap G) \oplus C_2' \oplus \dots \oplus C_n'$. Moreover, $C_1' = (C \oplus E_1') \cap G \subseteq (C \oplus C_1) \cap G = (G \cap C) \oplus C_1 = C_1$, since $G \cap C = 0$, as needed. Similarly, C has n-exchange.

(b) Assume that $G = A \oplus B = C_1 \oplus \dots \oplus C_n$ with $n > 2$. Then $G = A \oplus B = A \oplus E_1' \oplus C_n'$, where $E_1' \subseteq C_1 \oplus \dots \oplus C_{n-1}$ and $C_n' \subseteq C_n$, since A has 2-exchange. But $C_1 \oplus \dots \oplus C_{n-1} = (C_1 \oplus \dots \oplus C_{n-1}) \cap (A \oplus E_1' \oplus C_n') = E_1' \oplus E_1''$, where

$E'' = (C_1 \oplus \dots \oplus C_{n-1}) \cap (A \oplus C_n')$. Also $C_n = C_n \cap G = C_n \cap (E_1' \oplus A \oplus C_n') = C_n' \oplus C_n''$ where $C_n'' = C_n \cap (A \oplus E_1')$. Hence, $G = A \oplus E_1' \oplus C_n' = (C_1 \oplus \dots \oplus C_{n-1}) \oplus C_n = E_1' \oplus E_1'' \oplus C_n' \oplus C_n''$. Therefore, $A \simeq E_1'' \oplus C_n''$, A has n-1 exchange by induction on n, and so E_1'' has n-1 exchange by (a). Thus, $C_1 \oplus \dots \oplus C_{n-1} = E_1' \oplus E_1'' = E_1'' \oplus C_1' \oplus \dots \oplus C_{n-1}'$ for some $C_i' \subseteq C_i$. Also, $A \oplus C_n' = E_1'' \oplus E_1'''$ for some E_1''' since $E_1'' \subseteq A \oplus C_n'$ is a summand of G. Finally, $G = E_1' \oplus A \oplus C_n' = E_1' \oplus E_1'' \oplus E_1''' = C_1 \oplus \dots \oplus C_{n-1} \oplus E_1''' = E_1'' \oplus C_1' \oplus \dots \oplus C_{n-1}' \oplus E_1''' = A \oplus C_n' \oplus C_1' \oplus \dots \oplus C_{n-1}'$ and $C_i' \subseteq C_i$ for $1 \le i \le n$. ///

The following theorem is a characterization of an abelian group A with finite exchange in terms of E(A).

<u>Lemma 8.22</u>. Assume that $G = A \oplus B = C_1 \oplus C_2$ is an abelian group with associated orthogonal idempotents $\{a,b\}$ and $\{c_1, c_2\}$. Then $G = A \oplus C_1' \oplus C_2'$ with $C_i' \subseteq C_i$ iff there are $g, h \in \mathrm{Hom}(G,A)$ such that $gc_1g = g$ and $h(a-c_1a)(a-gc_1a) = a - gc_1a$. Furthermore, g and h may be chosen so that $g^2 = g$, $h^2 = h$, and $ha = hc_2a = a - gc_1a$.

<u>Proof</u>. ($\Rightarrow$) Write $G = A \oplus C_1' \oplus C_2'$ with $C_i' \subseteq C_i$ and with orthogonal idempotents $\{\alpha, c_1', c_2'\}$. Then $1_G = \alpha + c_1' + c_2'$ so that $c_1 = \alpha c_1 + c_1'c_1 + c_2'c_1$. Now $\alpha c_1 = c_1 - c_1'c_1 - c_2'c_1$ and $(\alpha c_1)^2 = \alpha c_1 - \alpha c_1c_1'c_1 - \alpha c_1c_2'c_1$. But $\alpha c_1c_1'c_1 = \alpha c_1'c_1 = 0$ since $C_1' \subseteq C_1$ and $c_1 = 1_{C_1}$. Also $\alpha c_1c_2'c_1 = 0$ since $C_2' \subseteq C_2 = \mathrm{Kernel}(c_1)$. Thus $(\alpha c_1)^2 = \alpha c_1$. Similarly, $(\alpha c_2)^2 = \alpha c_2$.

Define $g = \alpha c_1$ and $h = \alpha c_2$. Then $gc_1g = \alpha c_1\alpha c_1 = \alpha c_1 = g$. Furthermore, $h(a-c_1a) = \alpha c_2(a-c_1a) = \alpha c_2a = a - \alpha c_1a$ (since $1_G = c_1 + c_2$ implies that $a = \alpha a = c_1a + \alpha c_2a) = a - gc_1a$. Thus $h(a-c_1a)(a-gc_1a) = (a-gc_1a)^2 = a-gc_1a$ since $gc_1agc_1a = gc_1gc_1a = gc_1a$.

($\Leftarrow$) Let $\theta_1 = gc_1 : G \to A$. Then $\theta_1^2 = gc_1gc_1 = gc_1 = \theta_1$ is an idempotent of E(G). Define $\theta_2 = (1_G-\theta_1)h(1_G-c_1)(1_G-\theta_1)$. Then $\theta_1\theta_2 = \theta_2\theta_1 = 0$ and $\theta_2^2 = (1_G-\theta_1)h(1_G-c_1)(1_G-\theta_1)h(1_G-c_1)(1_G-\theta_1) = (1_G-\theta_1)h(a-c_1a)(a-gc_1a)h(1_G-c_1)(1_G-\theta_1)$ (since $g,h : G \to A$ and $a = 1_A) = (1_G-\theta_1)(a-gc_1a)h(1_G-c_1)(1_G-\theta_1) = (1_G-\theta_1)(1_G-gc_1)h(1_G-c_1)(1_G-\theta_1) = \theta_2$ since $\theta_1 = gc_1$ is idempotent.

Define $A_1 = \theta_1(G)$ and $A_2 = \theta_2(G)$. Then $A = A_1 \oplus A_2$ since $\theta_1 a + \theta_2 a = gc_1a + (a-gc_1a)h(a-c_1a)(a-gc_1a) = gc_1a + (a-gc_1a)^2 = gc_1a + (a-gc_1a)$ ($gc_1g = g$ implies that gc_1a is idempotent in $E(A)$) $= a = 1_A$ and θ_1, θ_2 are orthogonal idempotents in $E(G)$. Now $G = A_1 \oplus \text{Kernel}(\theta_1) = A_2 \oplus \text{Kernel}(\theta_2)$ with $A_2, C_2 \subseteq \text{Kernel}(\theta_1)$ since $\theta_1\theta_2 = 0 = \theta_1 c_2$. Moreover, $A_1, C_1' \subseteq \text{Kernel}(\theta_2)$ where $C_1' = C_1 \cap \text{Kernel}(\theta_1)$ since $\theta_2\theta_1 = 0$ and if $x \in C_1'$ then $(1_G-c_1)(1_G-\theta_1)(x) = (1_G-c_1)(x) = 0$. Therefore $\text{Kernel}(\theta_1) = \text{Kernel}(\theta_1) \cap (C_1 \oplus C_2) = C_2 \oplus C_1'$; $\text{Kernel}(\theta_2) \cap \text{Kernel}(\theta_1) = C_1' \oplus C_2'$, where $C_2' = \text{Kernel}(\theta_2) \cap C_2$; $\text{Kernel}(\theta_2) = (A_1 \oplus \text{Kernel}(\theta_1)) \cap \text{Kernel}(\theta_2) = A_1 \oplus (\text{Kernel}(\theta_1) \cap \text{Kernel}(\theta_2)) = A_1 \oplus C_1' \oplus C_2'$ and $G = A_2 \oplus \text{Kernel}(\theta_2) = A \oplus C_1' \oplus C_2'$ with $C_i' \subseteq C_i$, as needed. ///

<u>Theorem 8.23 (Monk [1])</u>: Let A be an abelian group. Then A has the finite exchange property iff whenever $f \in E(A)$ then there are $g, h \in E(A)$ with $gfg = g$ and $h(1_A-f)(1_A-gf) = 1_A-gf$.

<u>Proof</u>. (⇐) In view of Lemma 8.21, it suffices to prove that A has 2-exchange. Let $G = A \oplus B = C_1 \oplus C_2$ with orthogonal idempotents $\{a,b\}$ and $\{c_1,c_2\}$ and let $f = ac_1a \in E(A)$. Choose $g, h \in E(A)$ with $gfg = g$ and $h(1_A-f)(1_A-gf) = 1_A - gf$. Define $g' = ga : G \to A$ and $h' = ha : G \to A$. Then $g'c_1g' = gac_1aga = ga = g'$ and $h'(a-c_1a)(a-g'c_1a) = h(a-ac_1a)(a-gac_1a) = 1_A - gf = a - g'c_1a$. By Lemma 8.22, $G = A \oplus C_1' \oplus C_2'$ with $C_i' \subseteq C_i$ so that A has the 2-exchange property.

(⇒) Let $f \in E(A)$ and define $G = C_1 \oplus C_2$ to be the external direct sum of $C_1 = A$ and $C_2 = A$ with orthogonal idempotents $\{c_1, c_2\}$ and injections $e_1 : A \to G$, $e_2 : A \to G$ given by $e_1(x) = (x,0)$ and $e_2(x) = (0,x)$. Then $G = A' \oplus B$ with orthogonal idempotents $\{a', b\}$ where $A' = \{(f(x), x-f(x)) \mid x \in A\}$, $B = \{(x,-x) \mid x \in A\}$ and $a'(x,y) = (f(x+y), x+y-f(x+y))$.

Define $a_0 : G \to A$ by $a_0(x,y) = x + y$. Then $a_0a'e_1 = 1_A$, $a'e_1a_0 = a'$, $c_1a'e_1 = e_1f$, and $e_2(1_A-f)a_0 = a' - c_1a'$. Therefore, $a_0a' : A' \to A$ is an isomorphism with inverse $a'e_1 : A \to A'$.

Since $A' \cong A$ has the 2-exchange property there are $g', h' \in \text{Hom}(G,A')$ with $g'c_1g' = g'$ and $h'(a'-c_1a')(a'-g'c_1a') = a' - g'c_1a'$.

Define $g = a_0g'e_1 \in E(A)$. Then $gfg = a_0g'e_1fa_0g'e_1 = a_0g'c_1a'e_1a_0g'e_1 = a_0g'c_1a'g'e_1 = a_0g'c_1g'e_1 = a_0g'e_1 = g$. Define $h = a_0h'e_2 \in E(A)$. Then $h(1_A-f)(1_A-gf) = a_0h'e_2(1_A-f)(a_0a'e_1-a_0g'e_1f) = a_0h'e_2(1_A-f)\ a_0\ (a'e_1-g'c_1a'e_1) = a_0h'(a'-c_1a')(a'-g'c_1a')e_1 = a_0(a'-g'c_1a')e_1 = 1_A-gf$. ///

Corollary 8.24 (Warfield [6]): Suppose that A is an indecomposable abelian group. Then A has finite exchange iff E(A) is a local ring.

Proof. (⇐) Let $f \in E(A)$. If f is a unit of E(A) then $g = f^{-1}$ and $h = 0$ satisfies Theorem 8.23, since $gfg = g$ and $h(1-f)(1-gf) = 0 = 1-gf$. Otherwise, $1-f$ is a unit of E(A) in which case if $g = 0$ and $h = (1-f)^{-1}$ then $gfg = g = 0$ and $h(1-f)(1-gf) = 1-gf = 1$.

(⇒) Let $f \in E(A)$ and choose $g \in E(A)$ with $gfg = g$ and $h(1_A-f)(1_A-gf) = 1_A-gf$ (Theorem 8.23). Then $(gf)^2 = gf$ and $(fg)^2 = fg$. Since A is indecomposable, $fg = gf = 0$ or $fg = gf = 1$. If $gf = 1$ then f is a unit. If $gf = 0$ then $h(1-f)(1-gf) = h(1-f) = 1-gf = 1$ for some $h \in E(A)$ so that $1-f$ is a unit. Therefore, E(A) is a local ring. ///

Corollary 8.25. Let A be a finite rank torsion free group.

(a) If A has finite exchange then 1 is in the stable range of E(A).

(b) There is a rank-1 group A such that 1 is in the stable range of E(A) and A does not have finite exchange.

Proof. (a) In view of Lemma 8.21 and Corollary 8.7a it is sufficient to assume A indecomposable. Then E(A) is a local ring by Corollary 8.24. If $f_1g_1 + f_2g_2 = 1 \in E(A)$ then f_1g_1 or f_2g_2 is a unit of E(A). If f_1g_1 is a unit then so is f_1 and $f_1 + f_2 \cdot 0$ is a unit. If f_2g_2, hence f_2, is a unit then $f_1 + f_2\ (f_2^{-1}(1-f_1)) = 1$ is a unit of E(A).

(b) is Example 8.3.c, wherein 1 is in the stable range of $E(R) \simeq R$, R is not a local ring and R is indecomposable. Now apply Corollary 8.24. ///

As an application of Theorem 8.23 the following special case of a result due to Arnold-Lady [1] is proved. The full result will be proved in Theorem 9.11.

Theorem 8.26. Assume that A and B are finite rank torsion free, every $0 \neq f \in E(A)$ is a monomorphism and that A is not isomorphic to a quasi-summand of B.

(a) If $G = A \oplus B = C_1 \oplus C_2$ then $G = A \oplus C_1' \oplus C_2'$ for some $C_i' \subseteq C_i$.

(b) If $A \oplus B \simeq A \oplus C$ then $B \simeq C$.

Proof. (a) Write $G = A \oplus B = C_1 \oplus C_2$ with associated orthogonal idempotents $\{a, b\}$ and $\{c_1, c_2\}$. Then $c_1 = c_1 \cdot 1 \cdot c_1 = c_1 a c_1 + c_1 b c_1$ and $ac_1a = ac_1ac_1a + ac_1bc_1a$. But $ac_1bc_1a = 0$ otherwise A would be isomorphic to a quasi-summand of B since $0 \neq \alpha = (ac_1b)(bc_1a) \in E(A)$ is a monomorphism with $A/\alpha(A)$ finite. Thus, $ac_1a = ac_1ac_1a$.

Let $f = ac_1a$, so that $f^2 = f$, and let $g = f$, $h = 1_A$ in $E(A)$. Then $gfg = g$ and $h(1_A - f)(1_A - gf) = (1_A - f)(1_A - f) = 1_A - f$ so that the proof of Theorem 8.23 applies.

(b) Write $G = A \oplus B = A_1 \oplus C_2$ with $A_1 \simeq A$, $C_2 \simeq C$. By (a), $G = A \oplus B = A \oplus A_1' \oplus C_2'$ where $A_1' \subseteq A_1$, $C_2' \subseteq C_2$. Thus, $A_1 = A_1' \oplus A_1''$ and $C = C_2' \oplus C_2''$ for some A_2'' and C_2''. Since A is indecomposable, $A_1' = 0$ or $A_1'' = 0$. In the first case, $G = A \oplus B = A \oplus C_2'$ so that $B \simeq C_2'$, a summand of C, and $B \simeq C$ since $\operatorname{rank}(B) = \operatorname{rank}(C)$. If $A_1'' = 0$ then $G = A \oplus B = A \oplus A_1 \oplus C_2'$ so that $B \simeq A_1 \oplus C_2'$. But $A \oplus C_2' \simeq C$ since $G = A \oplus A_1 \oplus C_2' = A_1 \oplus C_2$ so that $B \simeq C$. ///

In summary, failure of cancellation for rank-1 groups in $\mathcal{A}$ is rare.

Corollary 8.27. Suppose that A, B, C are finite rank torsion free groups and $\operatorname{rank}(A) = 1$. If $A \oplus B \simeq A \oplus C$ then $B \simeq C$ in the following cases:

(a) $A \simeq Z$;

(b) 1 is in the stable range of $E(A)$;

(c) B has a summand isomorphic to A;

(d) For some $n > 1$, B^n has a summand isomorphic to A and B has self-cancellation;

(e) If B is nearly isomorphic to B' then $B \simeq B'$;

(f) A is not isomorphic to a quasi-summand of B.

Proof. (a), (b) Corollary 8.12.

(c), (d) Corollary 8.16, noting that rank-1 groups have self-cancellation by Corollary 8.17d.

(e) Corollary 7.17 implies that B is nearly isomorphic to C whence $B \simeq C$.

(f) Theorem 8.26. ///

On the other hand, a rank-1 group A and rank-2 groups B and C with $A \oplus B \simeq A \oplus C$ but $B \not\simeq C$ were constructed in Example 2.10. For such an example: (i) B is quasi-isomorphic to $A \oplus X$ for some rank-1 group X (8.27.f); (ii) type(A) and type(X) are incomparable (8.27.c and Theorem 2.3), (iii) $A \not\simeq Z$ and 1 is not in the stable range of E(A) (8.27.a and b), (iv) B and C are nearly isomorphic but not isomorphic (Corollary 7.17).

EXERCISES

8.1 Prove that if R is a ring then the following sets are equal.

(a) $\mathcal{J}(R)$;

(b) The intersection of the maximal left ideals of R ;

(c) $\{r \in R \mid 1-sr\}$ is a unit of R for all $s \in R$;

(d) $\{r \in R \mid 1-rs\}$ is a unit of R for all $s \in R$.

Conclude that $\mathcal{J}(R)$ is an ideal of R and show that $\mathcal{J}(R/\mathcal{J}(R)) = 0$.

8.2 Assume that R is a ring. Prove that

(a) If $R = R_1 \times \ldots \times R_k$ then $\mathcal{J}(R) = \mathcal{J}(R_1) \times \ldots \times \mathcal{J}(R_k)$

(b) If $R = \mathrm{Mat}_n(S)$ then $\mathcal{J}(R) = \mathrm{Mat}_n(\mathcal{J}(S))$.

8.3 Show that if E(A) is a subring of Q and if $pA \neq A$ for each prime p of Z then 1 is not in the stable range of E(A).

8.4 Let A be a finite rank torsion free group. Then $\mathrm{Hom}(A,Z) = 0$ iff for each $0 \neq n \in Z$ with $nA \neq A$ there is a finite rank torsion free group G with $E(G) \simeq Z$, $\mathrm{Hom}(A,G) = 0$, and an epimorphism $G \to A/nA$. (Hint: Consider Example 2.7).

§9. Subrings of Finite Dimensional Q-Algebras

Recall that rings have identities and that ring homomorphisms preserve identities. In particular, if R is a subring of S then $1_R = 1_S$.

A right (left) ideal I of a ring R is nilpotent if $I^{(n)} = I \cdot I \cdot \ldots \cdot I = 0$ for some $0 < n \in Z$. The nil radical of R, $N(R)$, is the right ideal of R generated by all nilpotent right ideals of R.

Proposition 9.1. (a) $N(R)$ is an ideal of R, $N(R) \subseteq J(R)$ and $N(R/N(R)) = 0$.

(b) If R is artinian and noetherian then $N(R) = J(R)$ is nilpotent.

(c) Assume that the additive group of R is finite rank torsion free and embed R as a subring of $QR = Q \otimes_Z R$ via $r \to 1 \otimes r$. Then $N(R)$ is a pure subgroup of R, $N(R)$ is a nilpotent ideal of R, $QN(R) = N(QR) = J(QR)$ and $N(R) = J(QR) \cap R$.

Proof. Suppose that I and J are nilpotent right ideals of R. Then RI, the left ideal of R generated by I, is nilpotent since $(RI)^{(n)} = RI^{(n)} = 0$ whenever $I^{(n)} = 0$. Furthermore, $I + J$ is nilpotent since if $I^{(n)} = J^{(n)} = 0$ then $(I+J)^{(2n)} \subseteq (RI+RJ)^{(2n)} = 0$. Consequently, $x \in N(R)$ iff xR is a nilpotent right ideal of R.

(a) If $x \in N(R)$ then xR is nilpotent. Thus $Rx \subseteq RxR \subseteq N(R)$ is nilpotent so that $N(R)$ is an ideal. Furthermore, $N(R) \subseteq J(R)$ since if $x \in N(R)$ then $1-xr$ is a unit of R for each $r \in R$. Finally, if $x + N(R) \in N(R/N(R))$ then $xR + N(R)$ is a nilpotent right ideal of $R/N(R)$ from which it follows that $x \in N(R)$. Thus, $N(R/N(R)) = 0$.

(b) From (a) $N(R) \subseteq J(R)$. Since $J(R) \supseteq \ldots \supseteq J(R)^{(n)} \supseteq 0 \ldots$ is a descending chain of ideals of R and R is artinian there is some $0 < n \in Z$ with $J(R)^{(n)} = J(R)J(R)^{(n)} = J(R)^{(n+1)}$. But R is also noetherian so $J(R)$ is finitely generated. Then $J(R)^{(n)} = 0$, by Nakayama's Lemma (Exercise 9.3), so that $J(R) \subseteq N(R) \subseteq J(R)$.

(c) To see that $N(R)$ is pure in R, note that if $0 \neq k \in Z$, $x \in R$ and $kx \in N(R)$ then kxR is a nilpotent right ideal of R so that xR is a nilpotent right ideal of R, since R is torsion free. Now $\mathcal{N}(R) \subseteq Q\mathcal{N}(R) \subseteq \mathcal{N}(QR) = \mathcal{J}(QR)$ and $\mathcal{N}(QR)$ is nilpotent since QR, being a finite dimensional Q-algebra, is artinian and noetherian. Therefore, $\mathcal{N}(QR) \cap R$ is nilpotent, hence contained in $\mathcal{N}(R)$. But $Q(\mathcal{N}(QR) \cap R) = \mathcal{N}(QR)$ so that $\mathcal{N}(R) = \mathcal{N}(QR) \cap R = \mathcal{J}(QR) \cap R$ and $Q\mathcal{N}(R) = \mathcal{N}(QR) = \mathcal{J}(QR)$. ///

Proposition 9.2. (a) If $R = R_1 \times \ldots \times R_k$ is a ring product then $\mathcal{N}(R) = \mathcal{N}(R_1) \times \ldots \times \mathcal{N}(R_k)$.

(b) $\mathcal{N}(\mathrm{Mat}_n(R)) = \mathrm{Mat}_n(\mathcal{N}(R))$, where $\mathrm{Mat}_n(R)$ is the ring of $n \times n$ matrices with entries in R.

(c) If $e^2 = e \in R$ then $\mathcal{N}(eRe) = e\mathcal{N}(R)e$.

Proof. (a) I is a right ideal of R iff $I = I_1 \times \ldots \times I_k$, where each I_i is a right ideal of R_i. Moreover, I is nilpotent iff each I_i is nilpotent.

(b) Let $S = \mathrm{Mat}_n(R)$, $M = (m_{ij}) \in \mathcal{N}(S)$ and choose a set $\{e_{ij} | 1 \leq i, j \leq n\}$ of matrix units of S such that $e_{ij}e_{k\ell} = \delta_{jk}e_{i\ell}$. For each i, j $m_{ij} \in e_{ii}Me_{jj}R \subseteq e_{ii}MS$ is a nilpotent right ideal of R so that $\mathcal{N}(S) \subseteq \mathrm{Mat}_n(\mathcal{N}(R))$. Conversely, if $M \in \mathrm{Mat}_n(\mathcal{N}(R))$ then MS must be a nilpotent right ideal of S, since $MS = e_{11}MS \oplus \ldots \oplus e_{nn}MS$ and each $e_{ii}MS$ is a nilpotent right ideal of S.

(c) Note that eRe is a ring with identity e and that nilpotent right ideals of eRe are of the form eIe, I a nilpotent right ideal of R. ///

Corollary 9.3. Assume that the additive group of a ring R is finite rank torsion free.

(a) If R has no zero divisors then $\mathcal{N}(R) = 0$.

(b) If R is reduced and semi-local ($pR = R$ for almost all primes p of Z) and if $n = \pi\{p | pR \neq R\}$ then $nR \subseteq \mathcal{J}(R) \subseteq R$.

(c) Suppose that $R \simeq E(A)$, where $A = B \oplus D$ with $B \neq 0$ and $\mathrm{Hom}(D,B) = 0$. Then $\mathrm{Hom}(B,D) \subseteq \mathcal{N}(E(A))$.

Proof. (a) Clear.

(b) Let $r \in R$. Then $1-nr$ is not a zero divisor in R, since R is reduced, i.e., $B = \{s \in R \mid (1-nr)s = 0\}$ is pure in R and $nB = B$. Thus $1-nr$ is a unit in QR (as in Proposition 6.1) say $(1-nr)s = s(1-nr) = m$ for some $s \in R$ and minimal $0 < m \in Z$. If p is a prime of Z dividing m then p does not divide n, else p divides s by the minimality of m. Thus, m is a unit of R and $1-nr$ is a unit of R for each $r \in R$, i.e. $n \in J(R)$.

(c) Regard $\mathrm{Hom}(B,D)$ as a left ideal of $E(A)$ so that if $f \in \mathrm{Hom}(B,D)$ then $f(B) \subseteq D$ and $f(D) = 0$. Then $(E(A)f)^2 A \subseteq (E(A)f)(D) = 0$ so that $(E(A)f)^2 = 0$ and $f \in N(E(A))$. ///

Remark: It should be emphasized that nilpotent elements of R need not be in $N(R)$. E.g. if S is a domain and $1 < n \in Z$ then $N(\mathrm{Mat}_n(S)) = 0$, by Corollary 9.3 and Proposition 9.2, while $\mathrm{Mat}_n(S)$ has non-zero nilpotent elements.

Theorem 9.4: Suppose that the additive group of a ring R is finite rank torsion free. Then R is left(right) Noetherian iff $N(R)$ is finitely generated as a left(right) ideal of R.

Proof. ($\Rightarrow$) Every left(right) ideal of a left(right) Noetherian ring is finitely generated.

($\Leftarrow$) The proof is by induction on k, where $N(R)^{(k)} = 0$.

Assume $k = 1$, i.e. $N(R) = 0$. Then $J(QR) = 0$, by Proposition 9.1.c. If I is a left ideal of R then $QI = QRe$ for some $e^2 = e \in QR$ since QI is a left ideal of the semi-simple artinian ring QR. Choose $0 \neq n \in Z$ with $ne = r \in I$. Then $Ir \subseteq Rr \subseteq I$ and $Ir = nI$ since if $x \in I$ then $xr = xne = nxe = nx$. Thus I/Rr is finite, since I/nI is finite, so that I is finitely generated.

Now assume $\mathcal{N}(R)^{(k)} = 0$, $k > 1$. Now $\mathcal{N}(R/\mathcal{N}(R)^{(k-1)}) = \mathcal{N}(R)/\mathcal{N}(R)^{(k-1)}$ so that $R/\mathcal{N}(R)^{(k-1)}$ is left Noetherian by induction on k. Also $\mathcal{N}(R)$ is a finitely generated $R/\mathcal{N}(R)^{(k-1)}$ module hence left Noetherian as an R-module. Thus $\mathcal{N}(R)$ and $R/\mathcal{N}(R)$ are left Noetherian R-modules from which it follows that R is a left Noetherian ring. The proof for the right module case is similar. ///

Theorem 9.5. Let R be a ring and let $\bar{R} = R/\mathcal{N}(R)$. If M is a right R-module let $\bar{M} = M/M\mathcal{N}(R)$ be a right $\bar{R}$-module.

(a) Suppose that $M, N \in P(R)$. Then M and N are R-isomorphic iff $\bar{M}$ and $\bar{N}$ are $\bar{R}$-isomorphic.

(b) If $e + \mathcal{N}(R)$ is an idempotent of $\bar{R}$ then there is an idempotent f of R with $f + \mathcal{N}(R) = e + \mathcal{N}(R)$ and $fe = ef$.

(c) If $R = M \oplus M'$ as R-modules and $\bar{M} = M_1 \oplus M_2$ as $\bar{R}$-modules then $M = N_1 \oplus N_2$ as R-modules, where $\bar{N}_1 = M_1$ and $\bar{N}_2 = M_2$.

(d) If F is a finitely generated free R-module and $\bar{F} = M_1 \oplus M_2$ then $F = N_1 \oplus N_2$ with $\bar{N}_1 = M_1$ and $\bar{N}_2 = M_2$.

Proof. (a) ($\Rightarrow$) is clear.

($\Leftarrow$) Let $\pi_M : M \to \bar{M}$ be the canonical map and let $\bar{f} : \bar{M} \to \bar{N}$ be an isomorphism with inverse $\bar{g}$. Since M and N are R-projective there is $f : M \to N$ with $\pi_N f = \bar{f}\pi_M$ and $g : N \to M$ with $\pi_M g = \bar{g}\pi_N$. Then $\pi_M gf = \bar{g}\pi_N f = \bar{g}\bar{f}\pi_M = \pi_M$ so that $gf(M) + M\mathcal{N}(R) = M$. By Exercise 9.3, $gf(M) = M$ since M is a finitely generated R-module. Similarly, $fg(N) = N$ so that $f : M \to N$ is an isomorphism.

(b) Let $x = e$, $y = 1-e$ so that $xy = e - e^2 \in \mathcal{N}(R)$. Now xy is nilpotent, say $(xy)^n = 0$. But $xy = yx$, so that $1 = x + y =$ $(x+y)^{2n-1} = (x^{2n-1} + a_1x^{2n-2}y + \dots + a_{n-1}x^ny^{n-1}) + (a_nx^{n-1}y^n + \dots + y^{2n-1}) =$ $f + (1-f)$. Then $f(1-f) = 0$ since $xy = yx$ and $(xy)^n = 0$, hence $f = f^2$. Finally, if $\bar{f} = f + \mathcal{N}(R)$ then $\bar{f} = \overline{x^{2n-1}} = \overline{e^{2n-1}} = \bar{e}$ and $fe = ef$ since $xy = yx$.

(c) Let $M = \alpha R$, $\alpha^2 = \alpha \in R$. Then $\alpha R\alpha$ is a ring with identity α and $0 \to \alpha N(R)\alpha \to \alpha R\alpha \to \alpha\bar{R}\alpha \to 0$ is exact. But $N(\alpha R\alpha) = \alpha N(R)\alpha$, $\alpha R\alpha = E_R(M)$ and $\alpha\bar{R}\alpha = E_{\bar{R}}(\bar{M})$ so $\bar{M} = M_1 \oplus M_2$ gives rise to an idempotent $\bar{e}^2 = \bar{e} \in \overline{\alpha R\alpha}$ with $\overline{eM} = M_1$. By (b), $\bar{e}$ can be lifted to $f^2 = f \in \alpha R\alpha$ with $\overline{fM} = M_1$ and $\overline{(1-f)}(\bar{M}) = M_2$. If $N_1 = f(M)$ and $N_2 = (1-f)(M)$ then $M = N_1 \oplus N_2$ with $\bar{N}_i = M_i$ as needed.

(d) If $F \simeq R^n$ then $E_R(F) \simeq Mat_n(R)$ and $E_{\bar{R}}(\bar{F}) \simeq Mat_n(\bar{R})$. Thus, there is an exact sequence $0 \to N(E_R(F)) \to E_R(F) \to E_{\bar{R}}(\bar{F}) \to 0$ since $N(Mat_n(R)) = Mat_n(N(R))$ by Proposition 9.2.c. Now $\bar{F} = M_1 \oplus M_2$ implies that $M_1 = \overline{eF}$, $M_2 = \overline{(1-e)}\bar{F}$ for some $\bar{e}^2 = \bar{e} \in E_{\bar{R}}(\bar{F})$. By (b), $\bar{e}$ lifts to $f^2 = f \in E_R(F)$. Let $N_1 = f(F)$, $N_2 = (1-f)(F)$ so that $F = N_1 \oplus N_2$ with $\bar{N}_1 = M_1$ and $\bar{N}_2 = M_2$. ///

Corollary 9.6. Let A be a torsion free group of finite rank.

(a) $H_A : P(A) \to P(E(A))$ defined by $H_A(B) = Hom(A,B)$, is a category equivalence and $\bar{*} : P(E(A)) \to P(\overline{E(A)})$, defined by $(\bar{*})(M) = \bar{M}$, induces a 1-1 correspondence of isomorphism classes which preserves direct sums.

(b) $H_A : QP(A) \to P(QE(A))$ defined by $H_A(B) = Q\otimes_Z Hom(A,B)$, is a category equivalence and $\bar{*} : P(QE(A)) \to P(\overline{QE(A)})$, defined by $(\bar{*})(M) = \bar{M}$, induces a 1-1 correspondence of isomorphism classes which preserves direct sums.

Proof. (a) Theorem 5.1 and Theorem 9.5.

(b) Corollary 7.22 and Theorem 9.5. ///

Note that $\overline{QE(A)} = QE(A)/N(QE(A)) = QE(A)/J(QE(A))$ is a semi-simple artinian algebra, hence a direct sum of minimal right ideals. Corollary 9.6.b has a group-theoretic interpretation for quasi-summands of A:

Corollary 9.7. Let A be a finite rank torsion free group.

(a) Suppose that B and C are quasi-summands of A. Then B and C are quasi-isomorphic iff $\overline{QHom(A,B)}$ and $\overline{QHom(A,C)}$ are isomorphic right ideals of $\overline{QE(A)}$.

(b) Suppose that $A/(B_1 \oplus \dots \oplus B_k)$ is finite. Then $\overline{QE(A)} = \overline{QHom(A,B_1)} \oplus \dots \oplus \overline{QHom(A,B_k)}$ as a direct sum of right ideals and $\overline{QHom(A,B_i)}$ is a minimal right ideal of $\overline{QE(A)}$ iff B_i is strongly indecomposable.

(c) Suppose that $\overline{QE(A)} = I_1 \oplus \dots \oplus I_k$ is a direct sum of right ideals. Then for $1 \leq i \leq k$ there are subgroups B_i of A such that $A/(B_1 \oplus \dots \oplus B_k)$ is finite and $QHom(A,B_i) = I_i$. Each B_i is strongly indecomposable iff I_i is a minimal right ideal of $\overline{QE(A)}$. ///

For a ring R let $C(R)$ denote the center of R. Then $C(R_1 \times \dots \times R_k) = C(R_1) \times \dots \times C(R_k)$, and $C(\mathrm{Mat}_n(R)) \simeq C(R)$. If the additive group of R is finite rank torsion free then $C(R)$ is a pure subgroup of R, $C(QR) = QC(R)$, $C(R) \cap N(R) = N(C(R))$ and there is an embedding $C(R)/N(C(R)) \to C(R/N(R))$. This embedding may be proper for if $R = \left\{\begin{pmatrix} x & 0 \\ y & z\end{pmatrix} \middle| x \in Z,\ y,\ z \in Q\right\}$ then $C(R) \simeq Z$ while $C(R/N(R)) \simeq C(Z \times Q) \simeq Z \times Q$. Furthermore, if QR is a simple algebra then $C(QR)$ is a field so that $C(R)$ is an integral domain.

Theorem 9.8. Assume that R is a ring with finite rank torsion free additive group. There is $0 \neq n \in Z$ and rings $R_i \subseteq QR_i$ such that $n(R_1 \times \dots \times R_k) \subseteq R/N(R) \subseteq R_1 \times \dots \times R_k$ and each QR_i is a simple Q-algebra. Moreover, $n(C_1 \times \dots \times C_k) \subseteq C(R/N(R)) \subseteq C_1 \times \dots \times C_k$ where each $C_i = C(R_i)$ is an integral domain and QC_i is an algebraic number field.

Proof. Since $N(R/N(R)) = 0$, it suffices to assume that $N(R) = 0$. Then $J(QR) = 0$ so that QR is a semi-simple artinian algebra. Write $QR = K_1 \times \dots \times K_k$, each K_i a simple Q-algebra. Then $1_R = (1_{K_1}, \dots, 1_{K_k})$. Define $R_i = R\cdot 1_{K_i}$, the projection of R into K_i. Since each $1_{K_i} \in QR$ there is $0 \neq n \in Z$ with $n\cdot 1_{K_i} \in R$ for each i. Thus $n(R_1 \times \dots \times R_k) \subseteq R \subseteq R_1 \times \dots \times R_k$ and $QR_i = K_i$. Clearly, $n(C(R_1)) \times \dots \times C(R_k)) \subseteq C(R) \subseteq C(R_1) \times \dots \times C(R_k)$ and $QC(R_i) = C(K_i)$ is an algebraic number field for each i. ///

Theorem 9.9 (Pierce [1]). Suppose that R is a subring of QR, a finite dimensional Q-algebra, and that $N(R) = 0$.

(a) R is finitely generated as a $C(R)$-module.

(b) If QR is a simple algebra then $C(R)$ is an integral domain and R is a finitely generated torsion free $C(R)$-module.

Proof. (a) In view of Theorem 9.8, there is $0 \neq n \in Z$ with $n(R_1 \times \ldots \times R_k) \subseteq R \subseteq R_1 \times \ldots \times R_k$ where each $R_i \subseteq QR_i$ is a simple algebra, $n(C(R_1) \times \ldots \times C(R_k)) \subseteq C(R) \subseteq C(R_1) \times \ldots \times C(R_k)$ and each $C(R_i)$ is an integral domain. Thus it suffices to prove (b), in which case R is finitely generated as a $C(R)$-module, observing that each R_i/nR_i and $C(R_i)/nC(R_i)$ is finite.

(b) Assume that QR is a simple Q-algebra. Then $F = C(QR)$ is an algebraic number field and $C(R)$ is an integral domain contained in F. Moreover, QR is a faithful irreducible left $QE_Z(R)$-module: faithfulness is clear and if M is a non-zero $QE_Z(R)$-submodule of QR then M is a two-sided ideal of QR since left and right multiplication by elements of QR are in $QE_Z(R)$. Since QR is a simple algebra, $M = QR$.

As a consequence of Exercise 9.1.g, $QE_Z(R) = E_D(QR)$ where $D = E_{QE_Z(R)}(QR)$ is a division algebra. Now if $f \in D$ then f is right multiplication by $f(1) \in QR$ since $x \in QR$ implies that $f(x) = f(x \cdot 1_{QR}) = xf(1_{QR})$. Thus, D is isomorphic to a subfield of $F = C(QR)$, via $f \to f(1)$, since if $x \in QR$ then $f(x) = f(x \cdot 1) = xf(1)$ and $f(x) = f(1 \cdot x) = f(1)x$.

Now D is a field and QR is a D-vector space so choose $r_1, \ldots, r_k \in R$, a D-basis of QR. Let $p_i : QR \to D$ be a projection map, where $p_i(\Sigma d_j r_j) = d_i$. Then $p_i \in E_D(QR) = QE_Z(R)$ so there is $0 \neq n \in Z$ with $np_i(R) \subseteq R$ for $1 \leq i \leq k$. Let $x = d_1r_1 + \ldots + d_kr_k \in R$. Then $nx = nd_1r_1 + \ldots + nd_kr_k = np_1(x)r_1 + \ldots + np_k(x)r_k \in (D \cap R)r_1 \oplus \ldots \oplus (D \cap R)r_k$. Therefore, $nR \subseteq (D \cap R)r_1 \oplus \ldots \oplus (D \cap R)r_k \subseteq R$. Since $D \cap R$ is noetherian (Theorem 9.4), $nR \simeq R$ is a finitely generated $D \cap R$-module. Consequently, R is a finitely generated $C(R)$-module since $D \cap R \subseteq C(R) = F \cap R$. Finally, R is $C(R)$-torsion free for if $0 \neq s \in C(R)$, $x \in R$, $sx = 0$ then $C(R)/C(R)s$ is finite (Exercise 9.4) so $0 \neq s's \in Z$ for some $s' \in C(R)$. Since R is Z-torsion free, x must be zero. ///

The results of this section apply to the case $R = E(A) \subseteq K = QR$, for A a finite rank torsion free group. If R and R' are subrings of a ring K then R and R' are quasi-equal if there are non-zero integers m and n with $mR' \subseteq R$ and $nR \subseteq R'$.

Theorem 9.10. Suppose that A is a finite rank torsion free group and that $nA \subseteq B \subseteq A$ for some $0 \neq n \in Z$ where $B = A_1^{n_1} \oplus \ldots \oplus A_m^{n_m}$, each A_i is strongly indecomposable, and A_i is quasi-isomorphic to A_j iff $i = j$. Let $R = E(A)$ and $T_i = E(A_i)$.

(a) $QR/\mathcal{J}(QR) = K_1 \times \ldots \times K_m$ where each $K_i = \mathrm{Mat}_{n_i}(D_i)$ is a simple Q-algebra and $D_i = QT_i/\mathcal{J}(QT_i)$ is a division algebra.

(b) $R/\mathcal{N}(R)$ is quasi-equal to $\Pi_i \mathrm{Mat}_{n_i}(T_i/\mathcal{N}(T_i))$.

(c) $\mathcal{C}(R/\mathcal{N}(R))$ is quasi-equal to $\Pi_i \mathcal{C}(T_i/\mathcal{N}(T_i))$

(d) $R/\mathcal{N}(R)$ is finitely generated as a $\mathcal{C}(R/\mathcal{N}(R))$-module.

Proof. Note that (c) follows from (b) by Theorem 9.8 and (d) is a consequence of Theorem 9.9. Moreover, $T_i/\mathcal{N}(T_i) \subseteq D_i = QT_i/\mathcal{J}(QT_i) = Q(T_i/\mathcal{N}(T_i))$. Since $T_i = E(A_i)$ and A_i is strongly indecomposable QT_i is a local artinian Q-algebra (Corollary 7.8). Thus $\mathcal{J}(QT_i)$ is the set of non-units of QT_i so that D_i is a division algebra. Furthermore, $K_i = \mathrm{Mat}_{n_i}(D_i)$ is a simple Q-algebra. Therefore, (a) follows from (b).

(b) Note that $E(A)$ is quasi-equal to $E(B)$ as subrings of $QE(A)$ whence $E(A)/\mathcal{N}(E(A))$ is quasi-equal to $E(B)/\mathcal{N}(E(B))$ as subrings of $QE(A)/\mathcal{J}QE(A))$. Thus, it is sufficient to prove that $E(B)/\mathcal{N}(E(B)) = \Pi_i \mathrm{Mat}_{n_i}(T_i/\mathcal{N}(T_i))$.

Represent $E(B)$ as a matrix $(\mathrm{Hom}(A_i^{n_i}, A_j^{n_j}))_{i,j}$. Let $(I = (\oplus_i\{\mathcal{N}E(A_i^{n_i})\}) \oplus (\oplus\{\mathrm{Hom}(A_i^{n_i}, A_j^{n_j}) \mid i \neq j\}) \subseteq E(B)$. It suffices to prove that I is an ideal of E(B) and $I \subseteq \mathcal{N}(E(B))$, in which case $E(B)/I \simeq \Pi_i(E(A_i^{n_i})/\mathcal{N}(E(A_i^{n_i})))$ so that $I = \mathcal{N}(E(B))$, since $\mathcal{N}(E(B)/I) = 0$, and $E(B)/\mathcal{N}(E(B)) \simeq \Pi_i(E(A_i^{n_i})/\mathcal{N}(E(A_i^{n_i}))) \simeq \Pi_i \mathrm{Mat}_{n_i}(E(A_i)/\mathcal{N}(E(A_i))) = \Pi_i \mathrm{Mat}_{n_i}(T_i/\mathcal{N}(T_i))$ as needed.

To show that I is an ideal of $QE(A)$ let $f \in \mathrm{Hom}(A_i^{n_i}, A_j^{n_j})$, $x \in N(E(A_k^{n_k})) \subseteq I$ and $y \in \mathrm{Hom}(A_r^{n_r}, A_s^{n_s}) \subseteq I$ where $r \neq s$. Then, $fx \in I$ except possibly for the case that $i = j = k$. In this case, $fx \in N(E(A_i^{n_i})) \subseteq I$. Also, $fy \in I$, except possibly for the case that $s = i$ and $r = j$. In this case, $y : A_j^{n_j} \to A_i^{n_i}$ and $f : A_i^{n_i} \to A_j^{n_j}$ induce maps $A_j \to A_i \to A_j$. If this latter composite is non-zero, it must be an element of $N(E(A_j))$. Otherwise the composite is a monomorphism, since A_j is strongly indecomposable, so that A_j is a quasi-summand of A_i contradicting the choice of the A_i's. Consequently, $fy \in \mathrm{Mat}_{n_j}(N(E(A_j)) = N(\mathrm{Mat}_{n_j}(E(A_j))) = N(E(A_j^{n_j})) \subseteq I$. Similarly, $xf \in I$ and $yf \in I$.

To show that $I \subseteq N(E(B))$ it suffices to prove that $I^{(\ell)} = 0$ for some ℓ. If $x \in I^{(\ell)}$ then x is the sum of elements which are the composition of ℓ morphisms $A_i^{n_i} \to A_j^{n_j}$ for $i \neq j$ and morphisms in $NE(A_i^{n_i})$. Choose k with $N(E(A_i^{n_i}))^{(k)} = 0$ for $1 \leq i \leq m$. Choose ℓ large enough so that any composition of ℓ morphisms, as described above, has some subscript repeated at least k times. If $A_i^{n_i} \to A_j^{n_j} \to \ldots \to A_i^{n_i}$ is a repetition of the subscript i then, as above, the composition must be in $N(E(A_i^{n_i}))$. Consequently, $I^{(\ell)} = 0$. ///

<u>Theorem 9.11 (Arnold-Lady [1])</u>: Assume that $G = A \oplus B = C_1 \oplus C_2$ is finite rank torsion free and that if A' and B' are strongly indecomposable quasi-summands of A and B, respectively, then A' and B' are not quasi-isomorphic.

(a) $G = A \oplus B = A \oplus C_1' \oplus C_2'$ for some $C_i' \subseteq C_i$.

(b) If $A \simeq C_1$ then $B \simeq C_2$.

<u>Proof.</u> (a) Let $\{a,b\}$ and $\{c_1,c_2\}$ be the orthogonal idempotents associated with $G = A \oplus B = C_1 \oplus C_2$. In view of Lemma 8.22 and the proof of Theorem 8.23 ($\Leftarrow$) it suffices to prove that if $f = ac_1a$ then there are $g,h \in E(A)$ with $gfg = g$ and $h(1_A-f)(1_A-gf) = 1_A-gf$. Now $c_1 = c_1 \cdot 1_G \cdot c_1 = c_1ac_1 + c_1bc_1$ and $f = ac_1a = ac_1ac_1a + ac_1bc_1a = f^2 + ac_1bc_1a$.

By Corollary 7.7 $a = e_1 + \ldots + e_n$ for some $e_i^2 = e_i \in QE(A)$ such that $e_ie_j = 0$ if $i \neq j$, $A_i = n_ie_i(A)$ is a strongly indecomposable subgroup of A for some $0 \neq n_i \in Z$ and $A/(A_1 \oplus \ldots \oplus A_n)$ is finite. Similarly,

$b = f_1 + \ldots + f_m$ for some $f_j^2 = f_j \in QE(B)$ with $f_i f_j = 0$ if $i \neq j$, $B_i = m_i f_i(B)$ is a strongly indecomposable subgroup of B for some $0 \neq m_i \in Z$, and $B/(B_1 \oplus \ldots \oplus B_m)$ finite. Then $ac_1bc_1a = (\Sigma e_i)\ c_1\ (\Sigma f_i)\ c_1\ (\Sigma e_i)$ where each $e_i c_1 f_j c_1 e_k : A_k \to B_j \to A_i$ is regarded as an element of $QE(G)$. If A_k and A_i are quasi-isomorphic and if $e_i c_1 f_j c_1 e_k$ is a monomorphism then A_k is quasi-isomorphic to B_j contradicting the hypotheses. Otherwise, by Theorem 9.10, each $e_i c_1 f_j c_1 e_k \in \mathcal{J}(QE(A))$ whence $ac_1bc_1a \in \mathcal{J}(QE(A)) \cap E(A) = NE(A)$.

Now $f + NE(A) = f^2 + NE(A)$ is an idempotent of $E(A)/NE(A)$. Theorem 9.5.b implies that there is $y^2 = y \in E(A)$ with $y + NE(A) = f + NE(A)$ and $fy = yf$. Write $f = y + x$ for $x \in NE(A)$ and define $g = (1+x)^{-1}y \in E(A)$, noting that $1 + x$ is a unit of $E(A)$ since x is nilpotent. Then $gfg = (1+x)^{-1}y(y+x)(1+x)^{-1}y = (1+x)^{-1}(y+yx)(1+x)^{-1}y = (1+x)^{-1}y^2 = g$. Note that $gf = (1+x)^{-1}yf = (1+x)^{-1}fy = (1+x)^{-1}(y+xy) = y$. Define $h = (1-x)^{-1} \in E(A)$. Then $h(1-f)(1-gf) = (1-x)^{-1}(1-y-x)(1-y) = (1-x)^{-1}(1-y-x+xy) = (1-x)^{-1}(1-x)(1-y)$ $= 1-y = 1-gf$ as needed.

(b) By (a) $G = A \oplus B = A \oplus C_1' \oplus C_2'$ where $C_i' \subseteq C_i$. Now $C_1 = C_1 \cap G = C_1' \oplus C_1''$ where $C_1'' = C_1 \cap (A \oplus C_2')$ and $C_2 = C_2 \cap G = C_2' \oplus C_2''$ where $C_2'' = C_2 \cap (A \oplus C_1')$. Furthermore, $B \simeq C_1' \oplus C_2'$ and $A \oplus C_1' \simeq C_1 \oplus C_2''$ since $A \oplus C_1' \oplus C_2' = C_1 \oplus C_2 = C_1 \oplus C_2' \oplus C_2''$.

By assumption $A \simeq C_1 = C_1' \oplus C_1''$ so that $C_1' = 0$ since C_1' is isomorphic to summands of both A and B. Thus $G = C_1 \oplus C_2 = C_1 \oplus C_2' \oplus C_2''$ and $G = A \oplus B \simeq C_1 \oplus C_2'$. Comparing ranks yields $C_2'' = 0$. Thus $B \simeq C_2' \simeq C_2$ as desired.

EXERCISES

9.1 This exercise is a summary of some of the classical Wedderburn-Artin theory of artinian rings. Assume that R is an artinian ring. Prove the following (in any order):

(a) If R is semi-simple and I is a (left, right) ideal of R then I is generated by an idempotent of R.

(b) If 0 and 1 are the only idempotents of R then R is a local ring with $\mathcal{J}(R)$ as the unique maximal ideal of R.

(c) If R is semi-simple and $R = f_1R + f_2R$ then $R = f_1R \oplus I$ for some right ideal I contained in f_2R.

(d) If R is semi-simple then $R = R_1 \times \dots \times R_k$ where each R_i is a simple algebra.

(e) If R is simple then $R \simeq \mathrm{Mat}_n(D)$ for some division algebra D and $\mathcal{C}(R) \simeq \mathcal{C}(D)$ is a field.

(f) If D is a finite division algebra then D is a field.

(g) (Jacobson-Wedderburn Density Theorem) If R has a faithful irreducible left module M then R is a simple algebra isomorphic to $E_D(M)$ where $D = E_R(M)$ is a division algebra.

9.2 Suppose that S is a subring of an algebraic number field and that M is a finitely generated S-module. Prove that M is torsion free as an abelian group iff M is torsion free as an S-module. In this case there is a finitely generated free S-module F such that $M \subseteq F$ and F/M is finite (9.4 may be useful).

9.3 Prove Nakayama's Lemma: Assume that R is a ring, M a finitely generated left (right) R-module, N a submodule of M and I a left (right) ideal of R. If $I \subseteq \mathcal{J}(R)$ and if $M = N + IM$ $(M = N + MI)$ then $M = N$.

9.4 Prove that if R is a subring of an algebraic number field and if I is a non-zero ideal of R then R/I is finite. Prove that QR is the quotient field of R.

§10. Orders in Finite Dimensional Simple Q-algebras

The following notation will be used in this section:

K - a simple finite dimensional Q-algebra;

R - a subring of K with QR = K;

F - center(K), an algebraic number field;

S - a subring of F, an integral domain.

The ring R is an S-order in K if R is an S-algebra which is finitely generated as an S-module. For example, R is a $\mathcal{C}(R)$-order in QR by Theorem 9.10. If R is an S-order and $\{x_1, \ldots, x_k\}$ is a maximal S-independent subset of R then $X = Sx_1 \oplus \ldots \oplus Sx_k \subseteq R$ and R/X is finite since R is a finitely generated S-module and S/Ss is finite for each $0 \neq s \in S$. Furthermore, R is torsion free as an S-module.

An S-order R in K is a maximal S-order in K if whenever $R \subseteq R' \subseteq K$ and R' is an S-order in K then R = R'.

The notion of maximal S-order is a generalization of the notion of integral closure for subrings of algebraic number fields.

An element x of K is integral over S if there is $0 < n \in Z$ and $s_{n-1}, \ldots, s_1, s_0 \in S$ with $x^n + s_{n-1}x^{n-1} + \ldots + s_0 = 0$.

Proposition 10.1. Let $x \in K$. The following are equivalent:

(a) x is integral over S;

(b) S[x] is a finitely generated S-module.

(c) x is an element of a subring of K which is finitely generated as an S-module.

Moreover, if $x, y \in K$ are integral over S and xy = yx then $x \pm y$ and xy are integral over S.

Proof. (a) ⟹ (b) If $x^n = -s_{n-1}x^{n-1} - \ldots - s_0$ then S[x] is generated as an S-module by $\{1, x, \ldots, x^{n-1}\}$.

(b) ⟹ (c) $x \in S[x]$ is finitely generated as an S-module.

(c) ⟹ (a) Assume $x \in B = S\beta_1 + \ldots + S\beta_n$, a subring of K generated as an S-module by $\{\beta_1, \ldots, \beta_n\}$. Then $x\beta_i = \Sigma s_{ij}\beta_j$ for $s_{ij} \in S$ so that $(x\delta_{ij} - s_{ij})(\beta_1, \ldots, \beta_n)^t = 0$. If $d = \det(x\delta_{ij} - s_{ij})$ then $d \in S$ and $d\beta_j = 0$ for each j so that $dB = 0$ and $d = 0$. But $d(X) = \det(X \cdot I - (s_{ij}))$ is a polynomial with coefficients in S, leading coefficient 1, and x is a root of $d(X)$. Thus, x is integral over S.

If $x, y \in K$ are integral then $S[x] = Su_1 + \ldots + Su_m$ and $S[y] = Sv_1 + \ldots + Sv_n$. If $xy = yx$ then any two elements of $S[x]$ and $S[y]$ commute so that $S[x,y] = \sum_i \sum_j Su_iv_j$. Thus, every element of $S[x,y]$ is integral over S by (c).

Example 10.2. Let $S = Z$, $K = \mathrm{Mat}_2(Q)$, $x = \begin{pmatrix} 0 & 1/2 \\ 0 & 0 \end{pmatrix}$, $y = \begin{pmatrix} 0 & 0 \\ 1/2 & 0 \end{pmatrix}$. Then $x^2 = 0 = y^2$ so that x and y are integral over Z. However, $x + y = \begin{pmatrix} 0 & 1/2 \\ 1/2 & 0 \end{pmatrix}$ is not integral over S, since $X^2 - 1/4$ is the minimum polynomial of $x + y$ over Q. Consequently, the hypothesis $xy = yx$ in Proposition 10.1 is necessary.

Note that QS, a subfield of F, is the quotient field of S. Define $\bar{S}$, the integral closure of S in F to be $\{x \in F \mid x$ is integral over $S\}$. Then $\bar{S}$ is an S-subalgebra of F, by Proposition 10.1, and $\bar{\bar{S}} = \bar{S}$. The ring S is integrally closed in F if $\bar{S} = S$ and is integrally closed if S is integrally closed in QS.

Corollary 10.3.

(a) S is a maximal S-order in QS iff S is integrally closed.

(b) If R is a (maximal) S-order in K and if $0 < n \in Z$ then $\mathrm{Mat}_n(R)$ is a (maximal) S-order in $\mathrm{Mat}_n(K)$.

Proof. (a) (⟹) Assume S is a maximal S-order in QS and let $x \in QS$ be integral over S. Then $S \subseteq S[x]$ is an S-order in QS (Proposition 10.1) so that $x \in S$ and $S = S[x]$, i.e. S is integrally closed in QS.

($\Leftarrow$) Assume S is integrally closed and that $S \subseteq S'$ is an S-order in QS, say $S' = Sx_1 + \ldots + Sx_n$. Then S' is finitely generated as an S-module and $S' = S[x_1, \ldots, x_n]$ so that $S' \subseteq \overline{S}$ by Proposition 10.1. Since $\overline{S} = S$ in QS, $S = S'$ and so S is a maximal S-order in QS.

(b) Since K is a simple algebra, $Mat_n(K)$ is a simple algebra and $C(Mat_n(K)) = C(K) = F$. Then $Mat_n(R)$ is an S-order in $Mat_n(K)$ since R a finitely generated S-module implies $Mat_n(R)$ is a finitely generated S-module, $QMat_n(R) = Mat_n(QR) = Mat_n(K)$, and S is a subring of $C(Mat_n(K))$.

Now assume R is a maximal S-order in K and that $Mat_n(R) \subseteq \Gamma' \subseteq Mat_n(K)$ where Γ' is an S-order in $Mat_n(K)$. Define $R' = \{x \in K | x$ occurs as the entry of some $\gamma' \in \Gamma'\}$. Since $R \subseteq R'$ it suffices to prove that R' is an S-order in K, in which case $R = R'$ and $Mat_n(R) = \Gamma'$.

For $\lambda \in R'$ define $E_\lambda = (m_{ij}) \in Mat_n(K)$ where $m_{11} = \lambda$ and $m_{ij} = 0$ otherwise. A routine computation shows that $E_\lambda \in Mat_n(R)\Gamma' Mat_n(R) \subseteq \Gamma'$. Also $E_\lambda + E_\mu = E_{\lambda+\mu}$, $E_\lambda E_\mu = E_{\lambda\mu}$, and $sE_\lambda = E_{s\lambda}$ for all $s \in S$. Thus $\{E_\lambda | \lambda \in R'\}$ is an S-subalgebra of Γ', R' is an S-subalgebra of K, and $\phi : R' \to \Gamma'$, given by $\phi(\lambda) = E_\lambda$ is an S-algebra monomorphism. But Γ' is a finitely generated module over the Noetherian domain S so that R' must be an S-order in K. ///

The next series of results is devoted to proving that if R is an S-order in K then $R \subseteq \overline{R}$, where $\overline{R}$ is a maximal S-order in K with $\overline{R}/R$ finite.

Assume that F' is a subfield of F. If $x \in K$ then μ_x, left multiplication by x, is an F'-linear transformation of K, regarded as an F'-vector space. Define $\text{min pol}_{K/F'}(x)$ to be the minimum polynomial of μ_x in $F'[X]$ and $\text{char pol}_{K/F'}(x)$ to be the characteristic polynomial of μ_x in $F'[X]$.

The following lemma contains well known results from elementary linear algebra.

<u>Lemma 10.4</u>. Let $x \in K$, $f(X) = \text{min pol}_{K/F'}(x)$ and $g(X) = \text{char pol}_{K/F'}(x)$

(a) $f(X)$ is the unique polynomial in $F'[X]$ with leading coefficient 1 and $f(x) = 0$.

(b) $g(x) = 0$,

(c) If $h(X) \in F'[X]$ with $h(x) = 0$ then $f(X)$ divides $h(X)$ in $F'[X]$.

(d) $f(X)$ divides $g(X)$ in $F'[X]$ and $f(X)$, $g(X)$ have the same irreducible factors in $F'[X]$.

(e) Let $(u_1, \ldots, u_n)$ be an F'-basis of K. Then $xu_j = \Sigma a_{ij}u_i$ for $a_{ij} \in F'$ and (a_{ij}) is the matrix of μ_x relative to $(u_1, \ldots, u_n)$. Hence $g(X) = \det(\delta_{ij}X - a_{ij})$ is independent of the choice of basis $(u_1, \ldots, u_n)$.

Note that $\min \mathrm{pol}_{K/F'}(x)$ does not depend on K but $\mathrm{char\ pol}_{K/F'}(x)$ does depend on K, having degree $= \dim_{F'}(K)$. Write $\mathrm{char\ pol}_{K/F'}(x) = X^m + f_{m-1}X^{m-1} + \ldots + f_0 \in F'[X]$ where $m = \dim_{F'}(K)$ and define $\underline{T_{K/F'}(x)} = -f_{m-1}$ and $\underline{N_{K/F'}} = (-1)^m f_0$, the trace and norm, respectively.

Lemma 10.5.

(a) $T_{K/F'} : K \to F'$ is an F'-linear transformation

(b) $N_{K/F'} : K \to F'$ and $N_{K/F'}(xy) = N_{K/F'}(x)N_{K/F'}(y)$, $N_{K/F'}(rx) = r^m N_{K/F'}(x)$ for all $x, y \in K$, $r \in F'$.

Proof. Note that $T_{K/F'}(x) = \mathrm{trace}(\mu_x)$ and $N_{K/F'}(x) = \det(\mu_x)$. The associated properties for the trace and determinant of matrices are well-known results of elementary linear algebra.

Theorem 10.6. An element x of K is integral over S iff $\min \mathrm{pol}_{K/F'}(x) \in \overline{S}[X]$ where $\overline{S}$ is the integral closure of S in $F' = QS$.

Proof. ($\Leftarrow$) Let $\min \mathrm{pol}_{K/F'}(x) = x^n + s_{n-1}x^{n-1} + \ldots + s_0 \in \overline{S}[X]$ and let $S' = S[s_{n-1}, \ldots, s_0] \subseteq \overline{S}$. By Theorem 10.1, S' is a finitely generated S-module. Moreover, x is integral over S' so $S'[x]$ is a finitely generated S'-module. Consequently, $S'[x]$ is a finitely generated S-module. Again, by Theorem 10.1, x is integral over S.

(⟹) Let $g(X) = \min \mathrm{pol}_{K/F'}(x) \in F'[X]$ and assume that $x \in K$ is integral over S. Then there is $f(X) \in S[X]$ with leading coefficient 1 such that $f(x) = 0$, whence $g(X)$ divides $f(X)$ in $F'[X]$, say $f(X) = g(X)h(X)$ for some $h(X) \in F'[X]$ with leading coefficient 1.

It is sufficient to prove that $g(X) \in \overline{S}[X]$. If S is a principal ideal domain, or a unique factorization domain, then the standard version of Gauss' Lemma would apply. However, Gauss's Lemma applies in a more general setting.

Let $L \supseteq F'$ be a splitting field for $f(X)$ and write $f(X) = \Pi(X-x_i)$ for $x_i \in L$. Let S_L be the integral closure of S in L. Then $f(X)$ has leading coefficient 1, $f(X) \in S_L[X]$, and $f(x_i) = 0$ for each i where each $x_i \in S_L$. Now $f(X) = g(X)h(X) = \Pi(X-x_i)$ so $g(X) = \Pi_I(X-x_i)$ and $h(X) = \Pi_J(X-x_i)$ for some index sets I and J. Thus $g(X), h(X) \in S_L \cap F'[X]$. Now $S \subseteq \overline{S} \subseteq S_L \cap F' \subseteq F' = QS$ and each element of $S_L \cap F'$ is integral over S. Consequently, $\overline{S} = S_L \cap F'$ and $g(X) \in \overline{S}[X]$. ///

Lemma 10.7. Write $K \simeq \mathrm{Mat}_m(D)$, where D is a division algebra with $D \supseteq F = C(D) = C(K)$. If E is a maximal subfield of D then $E \otimes_F K \simeq \mathrm{Mat}_n(E)$ where $n^2 = \dim_F K$.

Proof. First assume that $m = 1$ and $K = D$. Then D is a left $E\otimes_F D$ module, where $(x \otimes y)d = xdy$. In fact, D is a faithful irreducible $E\otimes_F D$ module since if $0 \neq d \in D$ then $(E\otimes_F D)d = D$ and D is faithful by Exercise 10.4(b). By the Jacobson density theorem $E\otimes_F D = E_L(D)$ where $L = E_{E\otimes_F D}(D)$ is a division algebra. Now $\phi : E \to L$ given by $\phi(x) =$ left multiplication by x is a ring monomorphism, noting that if $x \in E$, $x' \in E$, $y \in D$, $d \in D$ then $x(x' \otimes y)d = x(x'dy) = xx'dy = x'xdy = (x'\otimes y)(xd)$ so that ϕ is well defined. In fact, ϕ is onto since if $f \in L$ then f is left multiplication by $f(1)$ since $f(d) = f((1\otimes d)\cdot 1) = (1\otimes d)f(1) = f(1)d$ while $f(1)$ commutes with each element of E since $e(f(1)) = (e\otimes 1)f(1) = f(e\otimes 1) = f(e) = f(1)e$. Hence the subring of D generated by E and $f(1)$ is a field. But E is a maximal subfield of D so that $f(1) \in E$ and ϕ is onto. Thus $E\otimes_F D \simeq E_E(D) \simeq \mathrm{Mat}_n(E)$

where $n = \dim_E(D)$ and $n^2 = \dim_E(E\otimes_F D) = \dim_F(D)$, since if $\{x_i\}$ is an F-basis of D then $\{1 \otimes x_i\}$ is an E-basis of $E\otimes_F D$.

For the general case, $E\otimes_F K \simeq E\otimes_F \mathrm{Mat}_m(D) \simeq \mathrm{Mat}_m(E\otimes_F D) \simeq \mathrm{Mat}_{mn}(E)$ (Exercise 10.3) where $m^2n^2 = \dim_E(E\otimes_F K) = \dim_F K$. ///

Remark: In Lemma 10.7, E is called a splitting field for the simple algebra K.

Let $h : E\otimes_F K \to \mathrm{Mat}_n(E)$ be an isomorphism given by Lemma 10.7. For $x \in K$, define $tr_{K/F}(x)$, the reduced trace of x, to be trace$(h(1\otimes x))$ where $h(1\otimes x) \in \mathrm{Mat}_n(E)$. Note that $T_{K/F}(x) = \mathrm{trace}(\mu_x)$ from which it follows that $T_{K/F} = n(tr_{K/F})$. If $u_1, \ldots, u_m$ is an F-basis of K where $m = n^2$ then $\Delta_{K/F}(u_1, \ldots, u_m)$, the discriminant of $u_1, \ldots, u_m$, is defined to be $\det(tr_{K/F}(u_iu_j))$.

Lemma 10.8. If $u_1, \ldots, u_m$ is an F-basis of K then $\Delta_{K/F}(u_1, \ldots, u_m) \neq 0$.

Proof. Let $h : E\otimes_F K \to \mathrm{Mat}_n(E)$ be an isomorphism. Then $\{1\otimes u_i\}$ is an E-basis of $E\otimes_F K$ so that $\{h(1\otimes u_i)\}$ is an E-basis of $\mathrm{Mat}_n(E)$. Also $tr_{K/F}(u_iu_j) = \mathrm{trace}(h(1\otimes u_iu_j)$. Thus, it is sufficient to assume that $u_1, \ldots, u_m$ is an E-basis of $\mathrm{Mat}_n(E)$ and to prove that $\det(\mathrm{trace}(u_iu_j)) \neq 0$.

Suppose that $w_1, \ldots, w_m$ is another E-basis of $\mathrm{Mat}_n(E)$, say $w_i = \sum_k a_{i_k}u_k$. Then $w_iw_j = (\sum_k a_{i_k}u_k)(\sum_\ell a_{j_\ell}u_\ell) = \sum_\ell(\sum_k a_{i_k}u_ku_\ell)a_{j_\ell}$.

Since trace is E-linear $(\mathrm{trace}(w_iw_j)) = (a_{ik})(\mathrm{trace}(u_ku_\ell))(a_{\ell j})$ and $\det(\mathrm{trace}(w_iw_j) = \det(a_{ik})\det(\mathrm{trace}(u_ku_\ell))\det(a_{\ell j})$. Consequently, $\Delta_{K/F}(u_1,\ldots, u_m) \neq 0$ iff $\Delta_{K/F}(w_1, \ldots, w_m) \neq 0$.

In particular, $\{e_{ij}\}$ is an E-basis of $\mathrm{Mat}_n(E)$, where $e_{ij} = 1$ in i,j position and 0 elsewhere. Then $e_{ij}e_{k\ell} = \delta_{jk}e_{i\ell}$ so trace $(e_{ij}e_{k\ell}) = 0$ if $j \neq k$ or $i \neq \ell$ and 1 if $j = k$ and $i = \ell$. Thus, $(\mathrm{trace}(e_{ij}e_{k\ell}))$ has non-zero determinant, being a permutation matrix, so that $\Delta_{K/F}(u_1, \ldots, u_m) \neq 0$. ///

Theorem 10.9. Assume that $QS = F$ and that S is integrally closed (in F). Then R is an S-order in K iff each $x \in R$ is integral over S.

Proof. (=>) If R is an S-order then R is a finitely generated S-module. Thus each $x \in R$ is integral over S (Proposition 10.1).

(<=) Let $u_1, \ldots, u_m$ be an F-basis of K contained in R and let $\alpha = \det(T_{K/F}(u_iu_j))$. By Lemma 10.8, $\Delta_{K/F}(u_1, \ldots, u_m) = \det(tr_{K/F}(u_iu_j)) \neq 0$. But $T_{K/F}(u_iu_j) = n(tr_{K/F}(u_iu_j))$ for each i and j so that $\alpha \neq 0$.

In fact, $\alpha \in S$: each $u_iu_j \in R$ implies that u_iu_j is integral over S. Hence, $\min pol_{K/F}(u_iu_j) \in \overline{S}[X] = S[X]$, by Theorem 10.6 since $\overline{S} = S$. Furthermore, $\text{char pol}_{K/F}(u_iu_j) \in S[X]$ by Lemma 10.4.d, so that $T_{K/F}(u_iu_j) \in S$ for each i, j and $\alpha \in S$.

It is now sufficient to prove that $R \subseteq (1/\alpha)(Su_1 \oplus \ldots \oplus Su_m)$, in which case R is an S-order since S is a noetherian domain. Let $x = \Sigma r_iu_i \in R$ for some $r_i \in F$. Then $xu_j = \Sigma r_iu_iu_j$ and $T_{K/F}(xu_j) = \Sigma r_iT_{K/F}(u_iu_j)$. As above, each $T_{K/F}(xu_j) \in S$ so by Cramer's rule, each $r_i = s_i/\alpha$ for some $s_i \in S$. Thus $x \in (1/\alpha)(Su_1 \oplus \ldots \oplus Su_m)$ and $R \subseteq (1/\alpha)(Su_1 \oplus \ldots \oplus Su_m)$. ///

Corollary 10.10. Suppose that $QS = F$ and that S is integrally closed. If R is an S-order in K then there is a maximal S-order $\overline{R}$ in K with $R \subseteq \overline{R}$ and $\overline{R}/R$ finite.

Proof. Let $C = \{R' \subseteq K \mid R'$ is an S-order in K and $R' \supseteq R\}$. Then $R \in C$ is non-empty. If $R_1 \subseteq \ldots \subseteq R_\alpha \subseteq R_{\alpha+1} \subseteq \ldots$ is a chain of elements in C then $\overline{R} = \cup R_\alpha$ is in C as a consequence of Theorem 10.9. By Zorn's Lemma, C has a maximal element $\overline{R}$. ///

Observe that, as a consequence of the proof of Corollary 10.10, that if $R \subseteq R_1 \subseteq \ldots \subseteq R_n \subseteq R_{n+1} \subseteq \ldots$ is a chain of S-orders in K then there is n with $R_m = R_n$ for all $m \geq n$.

The next step is to remove the hypotheses in Corollary 10.10 that $QS = F$ and S is integrally closed.

Theorem 10.11. Let S' be the integral closure of S in F. If S is integrally closed then S' is an S-order in F.

Proof. Let $F' = QS$. Then $T_{F/F'} : F \to F'$. In fact $T_{F/F'} : S' \to S$ since if $x \in S'$ then x is integral over S so that min $\text{pol}_{F/F'}(x) \in S[X]$ (Theorem 10.6, noting that $\overline{S} = S$). Hence char $\text{pol}_{F/F'}(x) \in S[X]$, by Lemma 10.4, so that $T_{F/F'}(x) \in S$.

A routine calculation shows that $QS' = F$, since F is an algebraic extension of F'. Let $x_1, \ldots, x_n$ be an F' basis of F contained in S'. Define $\phi : F \to (F')^n$ by $\phi(x) = (T_{F/F'}(x_1x), \ldots, T_{F/F'}(x_nx))$. Then ϕ is an F'-homomorphism since $T_{F/F'}$ is F'-linear. If $x \in \text{Kernel}(\phi)$ then $T_{F/F'}(x_ix) = 0$ for all i so that $T_{F/F'}(yx) = 0$ for all $y \in F$. But $T_{F/F'}(x^{-1}x) = T_{F/F'}(1) \neq 0$ if $x \neq 0$. Thus $x = 0$ and ϕ is a monomorphism. Finally, $\phi : S' \to S^n$, noting that $T_{F/F'}(s') \in S$ for all $s' \in S'$, is monic so that S' is a finitely generated S-module since S is noetherian.

Define J_F to be the integral closure of Z in F, called the ring of algebraic integers of F. As a consequence of Theorem 10.11, $J_F \simeq Z^n$, as a group, where $n = \dim_Q F$, noting that Z-orders are finitely generated torsion free abelian groups, hence free.

Corollary 10.12. Assume that $QS = F$ and let $\overline{S}$ be the integral closure of S in F. Then $\overline{S} = SJ_F$, $\overline{S}$ is an S-order in F, and $\overline{S}/S$ is finite.

Proof. First of all, $S \subseteq SJ_F = \{\Sigma s_i j_i \mid s_i \in S, j_i \in J_F\} \subseteq F$. Since J_F is a finitely generated free abelian group, SJ_F is a finitely generated S-module. Hence $S \subseteq SJ_F \subseteq \overline{S}$ by Proposition 10.1. But $\overline{S}$ is integral over S hence integral over SJ_F and $J_F \subseteq SJ_F$. Since J_F is integrally closed in F, SJ_F is integrally closed in F (Exercise 10.1). Therefore, $\overline{S} = SJ_F$ whence $\overline{S}$ is an S-order in F and $\overline{S}/S$ is finite since if $\overline{S} = Sx_1 + \ldots + Sx_n$ there is $0 \neq m \in Z$ with $mx_i \in S$ for each i so that $m\overline{S} \subseteq S \subseteq \overline{S}$ and $\overline{S}/m\overline{S}$ is finite.

Corollary 10.13. If R is an S-order in K then there is a maximal S-order $\overline{R}$ in K with $R \subseteq \overline{R}$ and $\overline{R}/R$ finite. Moreover, $\overline{R}$ is a maximal S-order in K iff $\overline{R}$ is a maximal S'-order in K where S' is the integral closure of $\overline{S}$ in F and $\overline{S}$ is the integral closure of S in QS.

Proof. Let $\overline{S}$ be the integral closure of S in QS so that $\overline{S}/S$ is finite and $\overline{S}$ is an S-order in QS (Corollary 10.12). Let S' be the integral closure of $\overline{S}$ in F so that S' is an $\overline{S}$-order in F (Theorem 10.11) hence an S-order in F. Now $R \subseteq S'R \subseteq QR = K$ and $S'R/R$ is finite since R and S'R are both S-orders in K. But S'R is also an S'-order in K with $QS' = F$ and S' integrally closed in F. Thus $S'R \subseteq \overline{R}$ for some maximal S'-order $\overline{R}$ in K with $\overline{R}/S'R$ finite (Corollary 10.10). Hence $R \subseteq \overline{R}$, $\overline{R}$ is an S-order in K and $\overline{R}/R$ is finite. If $\overline{R} \subseteq R'$ is an S-order in K then $\overline{R} \subseteq R' \subseteq S'R'$ so $\overline{R} = R' = S'R'$ since S'R' is an S'-order in K and $\overline{R}$ is a maximal S'-order in K. Thus $\overline{R}$ is a maximal S-order in K. Conversely, if $\overline{R}$ is a maximal S-order in K then $\overline{R} = S'\overline{R}$, since S' is an S-order in F, and $\overline{R}$ is a maximal S'-order in K. ///

The following Corollary summarizes the results of Sections 9 and 10.

Corollary 10.14. Suppose that R is a subring of QR, a finite dimensional Q-algebra, i.e. the additive group of R is finite rank torsion free.

(a) $R/N(R)$ is Noetherian;

(b) There is $0 \neq n \in Z$ with $n(R_1 \times \ldots \times R_k) \subseteq R/N(R) \subseteq R_1 \times \ldots \times R_k$ where each QR_i is a simple Q-algebra, $R_i \subseteq QR_i$, and R_i is a $C(R_i)$-order in QR_i.

(c) There is $0 \neq m \in Z$ with $m\overline{R} \subseteq R/N(R) \subseteq \overline{R}$ where $\overline{R} = \overline{R}_1 \times \ldots \times \overline{R}_k$, each $\overline{R}_i$ is a maximal $C(R_i)(\overline{C(R_i)})$-order in the simple algebra QR_i and each $\overline{C(R_i)}/C(R_i)$ is finite, where $\overline{C(R_i)}$ is the integral closure of $C(R_i)$ in $QC(R_i) = C(QR_i)$, an algebraic number field.

EXERCISES

10.1 Suppose that S is an integrally closed subring of QS, an algebraic number field, and that $S \subseteq T \subseteq QS$ are subring inclusions.

(a) Prove that if P is a prime ideal of S then S_P, the localization of S at P, is a principal ideal domain with unique maximal ideal P.

(b) Prove that $S_P \otimes_S T$ is isomorphic to an integrally closed subring T_P of QS. Moreover, $T = \cap\{T_P | P \text{ is a prime ideal of } S\}$.

(c) Prove that T is integrally closed in QS.

10.2 Suppose that F is an algebraic number field and K is a finite dimensional F-algebra. Show that if K contains a set $\{e_{ij} | 1 \le i \le m, 1 \le j \le m\}$ of m^2 distinct non-zero elements such that $e_{ij}e_{k\ell} = 0$ if $j \ne k$ and $e_{ij}e_{j\ell} = e_{i\ell}$ for each $1 \le i, j, k, \ell \le m$ then L, the F submodule of K generated by $\{e_{ij}\}$, is a subalgebra of K isomorphic to $\mathrm{Mat}_m(F)$.

10.3 Assume that F is an algebraic number field, E a finite dimensional field extension of F and that K is a finite dimensional F-algebra.

(a) Prove that $E \otimes_F \mathrm{Mat}_m(K) \simeq \mathrm{Mat}_m(E \otimes_F K)$.

(b) Prove that if $E \otimes_F K \simeq \mathrm{Mat}_n(E)$ then $E \otimes_F \mathrm{Mat}_m(K) \simeq \mathrm{Mat}_{mn}(E)$ and $n^2 = \dim_F K$.

10.4 Assume that $F = C(D) \subseteq E \subseteq D$ where D is a division algebra and E is a subfield of D.

(a) Show that D is an $E \otimes_F D$-module where $(x \otimes y)d = xdy$.

(b) Show that if $u \in E \otimes_F D$ with $uD = 0$ then $u = 0$ (it suffices to assume that $u = e_1 \otimes d_1 + \ldots + e_n \otimes d_n$ where $\{e_1, \ldots, e_n\} \subseteq F$-basis of E and $d_i \ne 0$ for each i; reduce to the case that $d_1 = 1$; and induct on n to show that $((1 \otimes d)u - x(1 \otimes u))D = 0$ for each $d \in D$ implies that each $d_i \in F = C(D)$).

§11. Maximal Orders in Finite Dimensional Simple Q-algebras:

The following notation is assumed in this section:

K - finite dimensional simple Q-algebra,

F - Center (K), an algebraic number field,

S - a subring of F with $QS = F$,

R - an S-order in $K = QR$.

A proper ideal P of R is prime if whenever I and J are ideals of R with $IJ \subseteq P$ then $I \subseteq P$ or $J \subseteq P$. (Note that ideal means 2-sided ideal).

Theorem 11.1. Let P be a non-zero ideal of R. Then P is a prime ideal of R iff P is a maximal ideal of R.

Proof. (<=) If $IJ \subseteq P$ then $(I+P)(J+P) \subseteq P$ with $P \subseteq I + P$ and $P \subseteq J + P$. If $I \not\subseteq P$ and $J \not\subseteq P$ then $I + P = J + P = R$ by the maximality of P so that $R \subseteq P$, a contradiction.

(=>) Let $P' = P \cap S$, a proper prime ideal of S since $S \subseteq \mathcal{C}(R) = F \cap R$. Since $P \neq 0$ and $K = QR$ is a simple algebra, $K = QP$ and $P' \neq 0$. But S/P' is a finite integral domain (since $QS = F$ is the quotient field of S) and P' is a prime ideal of S, hence a field, so that P' is a maximal ideal of S. Also R/P is a finite dimensional S/P'-algebra, hence artinian. Moreover, $\mathcal{J}(R/P) = \mathcal{N}(R/P) = 0$, since P prime implies that 0 is the only nilpotent ideal of R/P. Finally, R/P is a simple algebra, in which case P is a maximal ideal of R, since otherwise $R/P = I_1/P \oplus I_2/P$ is the direct sum of ideals with $I_1I_2 \subseteq P$ and $I_i \not\subseteq P$ for $i = 1, 2$. ///

If M is a finitely generated S-submodule of K with $QM = K$ define $M^{-1} = \{x \in K | MxM \subseteq M\}$, $\mathcal{O}_\ell(M) = \{x \in K | xM \subseteq M\}$, and $\mathcal{O}_r(M) = \{x \in K | Mx \subseteq M\}$. Then $M^{-1} = \{x \in K | Mx \subseteq \mathcal{O}_\ell(M)\} = \{x \in K | xM \in \mathcal{O}_r(M)\}$. If M is a left (right) R-submodule of K then $R \subseteq \mathcal{O}_\ell(M)$ $(R \subseteq \mathcal{O}_r(M))$. Moreover, $\mathcal{O}_\ell(M)$ and $\mathcal{O}_r(M)$ are S-orders in K. Hence if R is a maximal S-order in K then $R = \mathcal{O}_\ell(M)$ $(R = \mathcal{O}_r(M))$ and $M^{-1} = \{x \in K | Mx \subseteq R\}$ $(M^{-1} = \{x \in K | xM \subseteq R\})$.

Lemma 11.2. Suppose that I is a non-zero ideal of R.

(a) There are non-zero prime ideals $P_1, P_2, \ldots, P_n$ of R with $I \supseteq P_1P_2 \ldots P_n$.

(b) If R is a maximal S-order in K and $I \neq R$ then $R \neq I^{-1}$.

Proof. (a) Assume that $\mathcal{S} = \{I | I$ is a non-zero ideal of R not containing a finite product of non-zero prime ideals of $R\}$ is non-empty. Since R is Noetherian, $\mathcal{S}$ has a maximal element M. Now M is not prime so there are ideals B and C of R with $BC \subseteq M$, $B \not\subseteq M$, $C \not\subseteq M$. Then $(B+M)(C+M) \subseteq M$ with $M \neq B + M$, $M \neq C + M$. By the maximality of M, $B+M$ and $C+M$ contain finite products of non-zero prime ideals a contradiction to $M \in \mathcal{S}$. Thus $\mathcal{S}$ is empty.

(b) Since I is an ideal of R and R is a maximal S-order in K, $I^{-1} = \{x \in K | xI \subseteq R\} = \{x \in K | Ix \subseteq R\}$. Assume $R = I^{-1}$ and embed I in M, a maximal ideal of R. Since $0 \neq I$ and $K = QR$ is simple, $K = QI = QM$ so that there is $0 \neq \alpha \in S \cap M \subseteq \mathcal{C}(R)$. By (a), $M \supseteq R\alpha \supseteq P_1 \ldots P_n$, each P_i a non-zero prime ideal of R and n minimal. Hence $M = P_i$ for some i, as a consequence of Theorem 11.1. Thus $M \supseteq R\alpha \supseteq BMC$, where B and C are products of the remaining prime ideals. Now $(1/\alpha)BMC \subseteq R$ so that $B(1/\alpha)MCB \subseteq B$ and $(1/\alpha)MCB \subseteq \mathcal{O}_r(B) = R$. Thus, $(1/\alpha)CB \subseteq M^{-1} \subseteq I^{-1} = R$, whence $CB \subseteq R\alpha$ contradicting the minimality of n, since BC is a product of $n-1$ non-zero prime ideals.

A left R-module M is an <u>R-generator</u> if $M^n \simeq R \oplus N$ as R-modules for some $0 < n \in Z$.

Theorem 11.3. The following are equivalent:

(a) R is a maximal S-order in K;

(b) If I is a non-zero ideal of R then $II^{-1} = I^{-1}I = R$;

(c) If I is a proper ideal of R then I is uniquely expressible as a product of prime ideals of R. If P_1, P_2 are prime ideals of R then $P_1P_2 = P_2P_1$.

(d) I is left (right) R-projective and a left (right) R-generator for each non-zero ideal I of R.

(e) If I is a non-zero ideal of R then I is a left (right) R-generator.

Proof. (a) ⟹ (b) Let $B = II^{-1}$, an ideal of R. Now $R \supseteq BB^{-1} = II^{-1}B^{-1}$ so $I^{-1}B^{-1} \subseteq I^{-1}$. Hence $R \subseteq B^{-1} \subseteq O_r(I^{-1}) = R$, since R is a maximal S-order. Thus $B^{-1} = R$ and $B = R$ by Lemma 11.2.b. Similarly, $I^{-1}I = R$.

(b) ⟹ (c) First of all, I is a product of non-zero prime ideals of R. Assume not and let M be a maximal counterexample. Then M is contained in a prime ideal P by Theorem 11.1. Thus $P^{-1}M \subseteq P^{-1}P = R$. Also $M \subsetneq P^{-1}M$, otherwise $R = MM^{-1} = P^{-1}MM^{-1} = P^{-1}$ and $P = RP = P^{-1}P = R$, which is impossible. By the maximality of M, $P^{-1}M$ is a product of prime ideals of R so that $M = PP^{-1}M$ is a product of prime ideals, a contradiction.

Let P and P' be prime ideals of R. Then $P' \supseteq P'P = PP^{-1}P'P$. Hence $P \subseteq P'$, in which case $P = P'$ and $PP' = P'P$, or else $P^{-1}P'P \subseteq P'$, in which case $P'P \subseteq PP'$ and symmetrically $PP' \subseteq P'P$. Consequently, $PP' = P'P$.

To show uniqueness, let $I = P_1P_2 \ldots P_n = P_1'P_2' \ldots P_m'$ be two products of prime ideals. Since prime ideals are maximal ideals, $P_i = P_j'$ for some i, j. By the preceding remarks, products of prime ideals commute so that $P_i^{-1}I = \Pi\{P_k \mid k \neq i\} = \pi\{P_k' \mid k \neq j\}$. Induction on n completes the proof.

(c) ⟹ (b) First assume that I is a non-zero prime ideal of R. Choose $0 \neq \alpha \in S \cap I \subseteq C(R)$ so that $R\alpha \subseteq I$. Write $R\alpha = P_1P_2 \ldots P_n$ as a product of prime ideals of R. Then some $P_i \subseteq I$ so that $P_i = I$ by Theorem 11.1. Now $(R\alpha)^{-1} = R(1/\alpha)$ so $(R\alpha)(R\alpha)^{-1} = (R\alpha)^{-1}R\alpha = R$. Also $P_i((R\alpha)^{-1} \pi_{j\neq i} P_j) = ((R\alpha)^{-1} \Pi_{j\neq i} P_j)P_i = R$ since products of prime ideals commute by (c). Therefore, it follows that $P_iP_i^{-1} = P_i^{-1}P_i = R$ whence $II^{-1} = I^{-1}I = R$. Finally, if $I = P_1P_2 \ldots P_n$ is a product of prime ideals then it follows that $II^{-1} = I^{-1}I = R$ since $I(P_n^{-1} \ldots P_1^{-1}) = (P_n^{-1} \ldots P_1^{-1})I = R$.

(b) ⟹ (d) Let I be a non-zero ideal of R with $II^{-1} = I^{-1}I = R$. To see that I is left projective write $1 = x_1y_1 + \ldots + x_ny_n$ for $x_i \in I^{-1}$, $y_i \in I$. Define $g : I \to R^n$ by $g(r) = (rx_1, \ldots, rx_n)$ and $f : R^n \to I$ by $f(r_1, \ldots, r_n) = r_1y_1 + \ldots + r_ny_n$. Then $fg(r) = rx_1y_1 + \ldots + rx_ny_n = r$ for all $r \in I$ so that $R^n \cong I \oplus I'$ as left R-modules.

Similarly, $II^{-1} = R$ implies that I is a left R-generator and $R = II^{-1} = I^{-1}I$ implies that I is a right R-projective generator.

(d) => (e) is clear.

(e) => (a) There is $0 \neq m \in Z$ with $I = m\overline{R} \subseteq R \subseteq \overline{R}$ where $\overline{R}$ is a maximal S-order in K (Corollary 10.13). Since I is an R-generator there is $f = (f_1, \ldots, f_n) : I^n \to R$ and $g = (g_1, \ldots, g_n) = R \to I^n$ with $f_1g_1(r) + \ldots + f_ng_n(r) = r$ for each $r \in R$. Now each f_i is right multiplication by some $x_i \in J = \{r \in K | I_r \subseteq R\}$ and each g_i is right multiplication by some $y_i = g_i(1) \in I$. Thus $1 = \Sigma y_i x_i$ so that $IJ = R$. Thus, $\overline{R} = \overline{R}R = \overline{R}IJ = IJ = R$ and R is a maximal S-order in K. The right generator case is similar.

An ideal I of S is <u>invertible</u> if $II^{-1} = S$. Define S to be a <u>Dedekind domain</u> if every non-zero ideal of S is invertible.

<u>Corollary 11.4</u>. The following are equivalent.

(a) S is integrally closed ;

(b) S is a Dedekind domain ;

(c) Every proper ideal of S is uniquely expressible as a product of prime ideals of S ;

(d) Every non-zero ideal of S is an S-projective generator ;

(e) Every non-zero ideal of S is an S-generator.

(f) Every non-zero ideal of S is projective.

<u>Proof.</u> Apply Corollary 10.3.a, Theorem 11.3, and Exercise 11.2.

<u>Corollary 11.5</u>. Suppose that I is a non-zero left (right) ideal of R and that R is a maximal S-order in K. Then I is a left (right) projective R-module.

<u>Proof.</u> Assume that I is a non-zero left ideal of R. Since QR = K is a simple algebra $K = QI \oplus M$. Then $I' = I \oplus (M \cap R)$ is a left ideal of R with $QI' = K$ and I is projective if I' is projective.

It now suffices to assume that QI = K. Then R/I is finite since R is a finitely generated S-module.

For an R-module M, define <u>proj dim</u>$_R$(M) ≤ 1 if there is an exact sequence of R-modules $0 \to L \to N \to M \to 0$ with L and N projective. Schanuel's Lemma implies that if proj $\dim_R(M) \leq 1$ and if $0 \to L' \to N' \to M \to 0$ is an exact sequence of R-modules with N' projective then L' is projective. Thus it suffices to prove that proj $\dim_R(R/I) \leq 1$.

Since R/I is finite there is a chain of R-modules $R/I = M_0 \supset M_1 \supset \ldots \supset M_t = 0$ with each M_i/M_{i+1} an irreducible R-module. If $0 \to M_{i+1} \to M_i \to M_i/M_{i+1} \to 0$ with proj $\dim_R(M_{i+1}) \leq 1$ and proj $\dim_R(M_i/M_{i+1}) \leq 1$ then proj $\dim_R(M_i) \leq 1$ (lift projective resolutions of M_{i+1} and M_i/M_{i+1} to a projective resolution of M_i). By induction on t, it suffices to assume that M is a finite irreducible R-module and prove that proj $\dim_R(M) \leq 1$.

Choose $0 \neq \ell \in Z$ with $\ell M = 0$. Then $\ell R = P_1 P_2 \ldots P_n$ is a product of prime ideals of R (Theorem 11.3). Since M is simple, $PM = 0$ for some prime ideal $P = P_i$ of R so that M is an R/P-module. But R/P is a simple algebra (Theorem 11.1) so that $M \oplus M' \simeq (R/P)^k$ for some $0 < k \in Z$. But P is projective by Theorem 11.3 so that proj $\dim_R(R/P) \leq 1$, hence proj $\dim_R(M) \leq 1$. ///

The preceding results demonstrate that maximal S-orders in K are non-commutative generalizations of Dedekind domains in algebraic number fields. Moreover, maximal S-orders in $F = QS$ are unique, namely the integral closure of S in F (Corollary 10.3). While maximal S-orders in K may be distinct, the following is true:

<u>Theorem 11.6</u>. Assume that R and R' are maximal S-orders in K. Then there is a category equivalence $P(R) \to P(R')$.

<u>Proof</u>. Let $B = RR' \subseteq K$. Then B is a finitely generated S-module (but not necessarily a subring of K) with $R \subseteq B$, $R' \subseteq B$. Hence B/R and B/R' are finite since $QR = QR' = K$ so that B is a finitely generated projective left R-module and a finitely generated projective right R'-module as a consequence of Corollary 11.5 (e.g. for some $0 \neq n \in Z$, $B \simeq nB$ is a left ideal of R).

Thus, $F : P(R) \to P(R')$, given by $F(M) = M \otimes_R B$ is a well-defined functor. Let $C = \{x \in K \mid Bx \subseteq R\}$, a finitely generated projective left R'-module and a finitely generated projective right R-module. Then $BC = R$ and $CB = R'$. Define $F' : P(R') \to P(R)$ by $F'(M') = M' \otimes_{R'} C$, a well-defined functor.

Now $B\otimes_{R'}C \to R = BC$ given by $b \otimes c \to bc$ is an isomorphism and $C\otimes_R B \to R' = CB$, given by $c \otimes b \to cb$, is an isomorphism, whence FF' is naturally equivalent to the identity on $P(R')$ and $F'F$ is naturally equivalent to the identity on $P(R)$. ///

For a prime p of Z let $S_p = Z_p \otimes_Z S$ and $R_p = Z_p \otimes_Z R$ so that $S \subseteq S_p \subseteq F = QS$ and $R \subseteq R_p \subseteq K = QR$ with $R = \cap R_p$ and $S = \cap S_p$.

<u>Theorem 11.7.</u> R is a maximal S-order in K iff R_p is a maximal S_p-order of K for each prime p of Z.

<u>Proof.</u> ($\Leftarrow$) If $R \subseteq R'$ is an S-order in K then $R_p \subseteq R'_p$ are S_p-orders in K for each p. Thus, $R_p = R'_p$ for each p so that $R = R'$.

($\Rightarrow$) Assume that $R_p \subseteq R'$ is a maximal S_p-order in K. Since R'/R_p is finite, there is $0 < i \in Z$ with $p^iR' \subseteq R_p \subseteq R'$. Let $T = p^iR' \cap R$, a finitely generated S-module contained in K with $T_p = p^iR'$. Then $R \subseteq O_\ell(T)$ and $O_\ell(T)$ is an S-order in K. Thus $R = O_\ell(T)$ whence $R_p = O_\ell(T)_p = O_\ell(T_p) = O_\ell(p^iR') = R'$. ///

Recall that R is semi-local iff $pR = R$ for almost all primes p of Z. For example, R_p is semi-local and S_p is a semi-local integral domain. If $S \neq F$ then S is reduced, as a group, since the divisible subgroup of S is an ideal of S and non-zero ideals have finite index in S. If $S \neq F$ is semi-local and $n = \pi\{p \mid pS \neq S\}$ then $nS \subseteq J(S) \subseteq S$ (Proposition 9.3.b). Moreover, if R is an S-order then R is semi-local; R is reduced as a group, and $nR \subseteq J(R) \subseteq R$.

<u>Theorem 11.8.</u> Assume that $S \neq F$ is semi-local and Dedekind and that R is a maximal S-order in K. Then S is a principal ideal domain and every non-zero left (right) ideal of R is principal. Furthermore, if R' is another maximal S-order in K then $R' = uRu^{-1}$ for some unit u of K.

<u>Proof.</u> As a consequence of Theorem 11.3 and the fact that in this case $R/J(R)$ is finite, hence artinian, $J(R) = M_1 \dots M_k$ is the product of distinct maximal ideals of R (since $N(R/J(R)) = J(R/J(R)) = 0$).

Now $R/J(R) \simeq R/M_1 \times \dots \times R/M_k$ as a consequence of the fact that $M_i + \pi\{M_j \mid j \neq i\} = R$, product of ideals commute by Theorem 11.3, and that if I and J are ideals of R with $I + J = R$ then $IJ = I \cap J$.

Let I be a non-zero left ideal of R. To show that I is R-principal, it is sufficient to prove that $I/\mathcal{J}(R)I \simeq Rx/\mathcal{J}(R)x$ for some $x \in R$; in which case an isomorphism $I/\mathcal{J}(R)I \to Rx/\mathcal{J}(R)x$ lifts to an isomorphism $I \to Rx$ by Nakayama's Lemma and the fact that I and Rx are R-projective (Corollary 11.5).

Note that $(R/\mathcal{J}(R)) \otimes_R I \simeq I/\mathcal{J}(R)I \simeq I/M_1I \oplus \ldots \oplus I/M_kI$ so it is sufficient to prove that $I/MI \simeq Rx/Mx$, where M is a maximal ideal of R, for some $x \in R$.

Now QI is a left ideal of the simple algebra $QR = K$ so that $QI = QRx$ for some $x \in I$. Since R is an S-order and S is noetherian, I and Rx are finitely generated S-modules hence I/Rx is finite. Now I/MI and Rx/Mx are R/M-modules and R/M is a simple artinian algebra so $I/MI \simeq L^k$ and $Rx/Mx \simeq L^m$, where L is an irreducible left ideal of R/M.

It suffices to prove that S is a principal ideal domain, in which case I and Rx are free S-modules with $S\text{-rank}(I) = S\text{-rank}(Rx)$ so that $\dim_T(I/MI) = k \dim_T(L) = m \dim_T(L)$, where $T = S/(M \cap S)$. Hence, $k = m$ and $I/MI \simeq Rx/Mx$ as needed.

Let J be a non-zero ideal of S so that $JJ^{-1} = S$ (Corollary 11.4). Write $1 = a_1b_1 + \ldots + a_sb_s$ with $a_i \in J$, $b_i \in J^{-1}$. Let $P_1, \ldots, P_t$ be the distinct maximal ideals of S recalling that $S/\mathcal{J}(S)$ is finite. By relabeling and allowing repetition if necessary, $a_ib_i \notin P_i$ for each i. Choose $u_i \in \cap \{P_j \mid j \neq i\} \setminus P_i$ and let $v = u_1b_1 + \ldots + u_nb_n \in J^{-1}$. Then Jv is an ideal of S and $Jv = S$ since $Jv \subseteq P_i$ is impossible for any i by the choice of u_i and a_ib_i. Thus $J = Sv^{-1}$ is principal.

Finally, let R' be another maximal S-order in K. Then $R \subseteq R'R$ and $R'R$ is a left R'-module, a right R-module, and a finitely generated S-module. Thus, there is $0 \neq n \in Z$ with $nR'R \subseteq R \subseteq R'R$. Since $nR'R$ is a right ideal of R, $rR = nR'R$ for some $r \in R$, as above. Moreover, $u = r/n$ is a unit of K since $(r/n)K = (r/n)QR = K$. But $uR = (r/n)R = R'R = R'R'R = R'uR \supseteq R'u$ so $R' \subseteq uRu^{-1}$. Furthermore, uRu^{-1} is an S-order in K and R' is a maximal S-order in K so that $R' = uRu^{-1}$. ///

Corollary 11.9. The following are equivalent:

(a) S is a Dedekind domain ;

(b) For each prime p of Z, S_p is a Dedekind domain ;

(c) For each prime p of Z, S_p is a principal ideal domain.

Proof. (a) ⟹ (b) is a consequence of Theorem 11.7, Corollary 11.4, and Corollary 10.3.a.

(b) ⟹ (c) follows from Theorem 11.8.

(c) ⟹ (a) Apply Theorem 11.7, Corollary 11.4, and Corollary 10.3.

Lemma 11.10 (Jordan-Zassenhaus). Suppose that $S = Z$ and that R is a Z-order in K. Then $\mathcal{I}(R) = \{(I) | I$ is a right ideal of $R\}$ is finite, where (I) is the isomorphism class of I.

Proof. First of all, if R' is another Z-order in $K = QR'$ and if $\mathcal{I}(R')$ is finite then $\mathcal{I}(R)$ is finite: Note that $R \cap R'$ is a Z-order in K with $Q(R \cap R') = K$ and that $nR \subseteq R \cap R' \subseteq R$ and $nR' \subseteq R \cap R' \subseteq R'$ for some $0 \neq n \in Z$ where R/nR and R'/nR' are finite. Thus, if $\mathcal{I}(R')$ is finite then $\mathcal{I}(R' \cap R)$, hence $\mathcal{I}(R)$, is finite noting that if I is a right ideal of R then nI is a right ideal of $R \cap R'$ and $n^2 I$ is a right ideal of $R \cap R'$.

Write $K = \mathrm{Mat}_m(D)$ for some division algebra D. If T is a Z-order in D then $\mathrm{Mat}_m(T)$ is a Z-order in K (Corollary 10.3). Moreover, if $\mathcal{I}(T)$ is finite then $\mathcal{I}(\mathrm{Mat}_m(T))$ is finite (Exercise 11.1). Thus, it suffices to assume that K is a division algebra and R is a Z-order in K.

Write $R = Zx_1 \oplus \dots \oplus Zx_n$ and $x_i x_j = \sum_k m_{ijk} x_k$ for $m_{ijk} \in Z$. If $x = \sum_i \lambda_i x_i \in R$ with $\lambda_i \in Z$ then $xx_j = \sum_i \sum_k \lambda_i m_{ijk} x_k$. Recall that $\mu_x : R \to R$ is left multiplication by x and $N_{K/Q}(x) = \det(\mu_x) = \det(M)$ where $M = (\sum_i \lambda_i m_{ijk})_{j,k}$ (Section 10). Diagonalize M using row operations and column interchanges to see that if $0 \neq x \in R$ then $\mathrm{card}(R/xR) = |N_{K/Q}(x)|$. Now $\det(M)$ is a homogeneous polynomial of degree n in $\lambda_1, \dots, \lambda_n$ with coefficients in Z. Hence there is $0 < c \in Z$, depending only on $\{m_{ijk}\}$ such that if $x = \sum \lambda_i x_i$ and $|\lambda_i| \leq t$ for $1 \leq i \leq n$ then $\mathrm{card}(R/xR) = |N_{K/Q}(x)| \leq c \cdot t^n$.

Let I be a proper right ideal of R so that R/I is finite since $K = QR$ is a division algebra. Choose $0 < s \in Z$ with $s^n \le \text{card}(R/I) < (s+1)^n$. Now if $S = \{x = \lambda_1 x_1 + \dots + \lambda_n x_n \mid \lambda_i \in Z,\ |\lambda_i| \le s+1\}$ then $\text{card}(S) \ge (s+1)^n > \text{card}(R/I)$. Thus there is $0 \ne x = \lambda_1 x_1 + \dots + \lambda_n x_n \in I$ with $|\lambda_i| \le 2(s+1)$ for each i (e.g. there are distinct elements s_1 and s_2 of S with $s_1 - s_2 = x \in I$). Hence $xR \subseteq I$ and $\text{card}(R/xR) = |N_{K/Q}(x)| \le c(2(s+1))^n$ while $\text{card}(R/I) \ge s^n$. Therefore, $\text{card}(I/xR) \le c2^n(s+1)^n/s^n = c2^n(1+1/s)^n \le c4^n$.

Finally, let $b = c4^n$. If I is a right ideal of R then there is $x \in I$ with $\text{card}(I/xR) \le b$. Thus, $bxR \subseteq bI \subseteq xR$ whence $bR \subseteq bx^{-1}I \subseteq R$ with $bx^{-1}I \simeq I$. Since R/bR is finite, $I(R)$ must be finite. ///

Let (B) denote the isomorphism class of a group B.

Theorem 11.11 (Lady [1]). Assume that A is a finite rank torsion free group.

(a) $\{(B) \mid B$ is a summand of $A\}$ is finite.

(b) $\{(B) \mid B$ is a finite rank torsion free group nearly isomorphic to $A\}$ is finite.

Proof. (a) In view of Corollary 9.6 there is a 1-1 correspondence from isomorphism classes of summands of A to isomorphism classes of right $E(A)/N(E(A))$-summands of $E(A)/N(E(A))$.

Thus it suffices to prove that if $R \subseteq QR$, a finite dimensional semi-simple Q-algebra then R has, up to isomorphism, only finitely many right R-summands. Choose $0 \ne n \in Z$ with $n\overline{R} \subseteq R \subseteq \overline{R}$, where $\overline{R} = R_1 \times \dots \times R_k$ and each R_i is a maximal $C(R_i)$-order in the simple algebra QR_i (Theorem 9.9). Thus, $\overline{R}$ has, up to isomorphism, only finitely many right $\overline{R}$-summands iff R has, up to isomorphism, only finitely many right R-summands, since $\overline{R}/R$ is finite. Therefore, since $\overline{R} = R_1 \times \dots \times R_k$, it suffices to assume QR is a simple algebra.

For each i, let $y_1 = 1, \dots, y_m$ be a Q-basis of the simple algebra QR. Then $y_\ell y_j = \Sigma q_{\ell jk} y_k$ for $q_{\ell jk} \in Q$. Clear denominators of $\{q_{\ell jk}\}$ to get a Q-basis $x_1 = 1, \dots, x_m$ of QR such that $T = Z\cdot 1 \oplus Z\cdot x_2 \oplus \dots \oplus Z\cdot x_m$ is a Z-order in $QR = QT$. Then, $T \cap R$ is a Z-order in QR with $QT = QR$ so it suffices to assume that T is a Z-order in QR with $T \subseteq R$ and $QT = QR$.

Let M be a right R-summand of R, say $M = eR$ for some $e^2 = e \in R$. Then $me \in T$ for some $0 \neq m \in Z$. By Lemma 11.10, $\mathcal{I}(T) = \{(I_1), \ldots, (I_k)\}$ is finite. Since meT is a right ideal of T, $meT \simeq I_j$ for some j. Thus, $mM = meR = (meT)R \simeq I_jR$ so that $M \simeq mM \simeq I_jR$ as right R-modules. Thus $\{(I_1R), \ldots, (I_kR)\}$ contains a complete set of representatives for isomorphism classes of right R-module summands of R.

(b) If B is nearly isomorphic to A then $B \oplus B' \simeq A \oplus A$ for some B' (Lemma 7.20). Now apply (a) to $A \oplus A$. ///

EXERCISES

11.1 Suppose that T is a Z-order in the finite dimensional Q-division algebra D. Prove that if $\mathcal{I}(T)$ is finite then $\mathcal{I}(\mathrm{Mat}_m(T))$ is finite.

11.2 Assume that R is an S-order in the simple algebra $QR = K$ where $S = \mathrm{Center}(R)$. Let $0 \neq m \in Z$ and $I = m\overline{R} \subseteq R \subseteq \overline{R}$ where $\overline{R}$ is a maximal S-order in K, and $J = \{r \in K \mid Ir \subseteq R\}$.

(a) Prove that if every non-zero ideal of R is projective then $R \subseteq IJ$ and IJ is an S-order in K.

(b) There is an example of a left hereditary R that is not a maximal S-order in K (Warfield-Huber).

11.3 Prove that if S is a subring of an algebraic number field and if every non-zero ideal of S is projective then S is integrally closed (i.e., S is a maximal S-order in QS). (See Exercise 11.2.)

§12. Near Isomorphism and Genus Class

Notation for this section is:

K - a semi-simple finite dimensional Q-algebra with $K = K_1 \times \ldots \times K_k$ where each K_i is a simple Q-algebra

F - Center(K) with $F = F_1 \times \ldots \times F_k$ where each $F_i = \text{Center}(K_i)$ is an algebraic number field.

R - a subring of K with $QR = K$

S - Center(R) so that $QS = F$

$\overline{R}$ - a subring of K containing R such that $\overline{R}/R$ is finite, $\overline{R} = R_1 \times \ldots \times R_k$, for each i, $QR_i = K_i$, R_i is a maximal S_i-order in K, and S_i is the integral closure of Center(R_i) in F_i so that S_i is a Dedekind domain.

Recall that R is a finitely generated S-module (Theorem 9.10) and that R is noetherian (Theorem 9.4). A finitely generated left(right) R-module M is a left(right) R-lattice if M is torsion free as an abelian group, necessarily of finite rank. Note that each left(right) ideal of R is a left(right) R-lattice.

Proposition 12.1. Assume that $R = \overline{R}$. Then each R-lattice is R-projective.

Proof. If M is an R-lattice then $M = M_1 \oplus \ldots \oplus M_k$ where each $M_i = R_iM$ is an R_i-lattice. It suffices to assume that QR is a simple algebra, since if each M_i is R_i-projective then M is R-projective.

Assume that QR is a simple algebra. Then QM is a $K = QR$-module so $QM \oplus M' \simeq K^n$ for some n. Thus, there are free R-modules F_1, F_2 with $F_1 \subseteq M \oplus F_2 \subseteq QM \oplus M'$ and $QM \oplus QF_2 = QF_1 = QM \oplus M'$. It suffices to assume $QM = K^n$, since if $M \oplus F_2$ is R-projective then M is R-projective. Since $QM = QR^n$ and R^n, M are finitely generated S-modules M is isomorphic to an R-submodule of R^n. But R is a maximal S-order in the simple algebra K so that R is left and right hereditary (Theorem 11.3). Therefore, M is R-projective since if R is left and right hereditary then submodules of projective R-modules are projective (Exercise 5.5). ///

Two R-lattices M and N are <u>in the same genus class</u> if M_p is R_p-isomorphic to N_p for each prime p of Z, where $M_p = Z_p \otimes_Z M$ is an R_p-lattice. Define M to be a <u>genus summand of N</u> if for each prime p of Z, M_p is isomorphic to an R_p-summand of N_p.

<u>Theorem 12.2.</u> Suppose that M and N are R-lattices.

(a) M is a genus summand of N iff for each $n \in Z$ there is an integer relatively prime to n and $f \in \mathrm{Hom}_R(M,N)$, $g \in \mathrm{Hom}_R(N,M)$ with $gf = m \cdot 1_M$

(b) M and N are in the same genus class iff for each $n \in Z$ there is an integer m relatively prime to n and $f \in \mathrm{Hom}_R(M,N)$, $g \in \mathrm{Hom}_R(N,M)$ with $gf = m \cdot 1_M$ and $fg = m \cdot 1_N$.

<u>Proof.</u> (a) ($\Leftarrow$) Given a prime p choose an integer m relatively prime to p and $f : M \to N$, $g : N \to M$ with $gf = m$. Then $f_p = 1 \otimes f : M_p \to N_p$ and $g_p = 1 \otimes g : N_p \to M_p$ with $g_p f_p = m$. Since m is a unit of Z_p, hence R_p, M_p is isomorphic to a summand of N_p.

($\Rightarrow$) If $R = K$ the result is clear. If $R \neq K$ then the maximal divisible subgroup $d(R)$ of R is an ideal of R. By passing to $R/d(R)$ it is enough to assume that R, hence M and N, are reduced as groups.

As in the proof of Theorem 7.16 (c)$\Rightarrow$(d) there is $0 \neq k \in Z$ with $kX \neq X$ for all non-zero subgroups X of R. Replacing n by the product of the distinct prime divisors of nk, it is sufficient to assume that $n = p_1 p_2 \cdots p_\ell$ is a product of distinct primes of Z with $nX \neq X$ for all non-zero subgroups X of R.

Define $R_n = \cap \{R_{p_i} \mid 1 \leq i \leq \ell\}$, $M_n = \cap \{M_{p_i} \mid 1 \leq i \leq \ell\}$ and $N_n = \cap \{N_{p_i} \mid 1 \leq i \leq \ell\}$. Then M_n and N_n are R_n-lattices with $(M_n)_{p_i} = M_{p_i}$ and $(N_n)_{p_i} = N_{p_i}$ for each $1 \leq i \leq \ell$. Moreover, R_n is reduced, by the choice of n, and R_n is semi-local so $nR_n \subseteq J(R_n) \subseteq R_n$ (Corollary 9.3).

For each $1 \leq i \leq \ell$, there is $f_i : M_{p_i} \to N_{p_i}$ and $g_i : N_{p_i} \to M_{p_i}$ with $g_i f_i = 1$. Since M_n and N_n are finitely generated R_n-modules it suffices to assume that $f_i : M_n \to N_n$ and $g_i : N_n \to M_n$ (multiply by an appropriate integer prime to p_i). Choose $\beta_i \in Z$ such that $\beta_i \equiv 1 (\mathrm{mod}\ p_i)$ and $\beta_i \equiv 0 (\mathrm{mod}\ p_j)$

if $j \neq i$ and define $f = \beta_1 f_1 + \ldots + \beta_\ell f_\ell : M_n \to N_n$, $g = \beta_1 g_1 + \ldots + \beta_\ell g_\ell : N_n \to M_n$. Now $M_n/nM_n \simeq (M_{p_1}/p_1 M_{p_1}) \oplus \ldots \oplus (M_{p_\ell}/p_\ell M_{p_\ell})$ and $N_n/nN_n \simeq (N_{p_1}/p_1 N_{p_1}) \oplus \ldots \oplus (M_{p_\ell}/p_\ell N_{p_\ell})$ and $\overline{f} : M_n/nM_n \to N_n/nN_n$ corresponds to $\overline{f}_1 \oplus \ldots \oplus \overline{f}_\ell$ while $g : N_n/nN_n$ corresponds to $\overline{g}_1 \oplus \ldots \oplus \overline{g}_\ell$. Each $g_i f_i = 1$, whence $gf(M_n) + nM_n = M_n$ and $gf(M_n) = M_n$, by Nakayama's Lemma, since $nR_n \subseteq J(R_n)$ and M_n is a finitely generated R_n-module. Since gf is an automorphism of M_n and M, N are finitely generated R-modules there is an integer relatively prime to n and $f' \in \mathrm{Hom}_R(M,N)$, $g' \in \mathrm{Hom}_R(N,M)$ with $g'f' = m$, as needed.

The proof of (b) is similar to the proof of (a). ///

Corollary 12.3. Assume that M and N are R-lattices and that R is semi-local. Then M and N are in the same genus class iff M and N are isomorphic.

Proof. (<=) Clear

(=>) Let $n = \pi\{p \mid pR \neq R\}$ and apply Theorem 12.2.b. ///

Corollary 12.4. Assume that M and N are R-lattices.

(a) If $R = \overline{R}$ then M and N are in the same genus class iff QM and QN are isomorphic as $K = QR$-modules.

(b) Let $n > 0$ be the smallest integer with $n\overline{R} \subseteq R \subseteq \overline{R}$ and let π_R be the set of prime divisors of n. Then M and N are in the same genus class iff $QM \simeq QN$ as K-modules and $M_p \simeq N_p$ as R_p-modules for each prime $p \in \pi_R$.

Proof. (a) (=>) If $M_p \simeq N_p$ then $QM \simeq QN$.

(<=) Write $M = M_1 \oplus \ldots \oplus M_k$ and $N = N_1 \oplus \ldots \oplus N_k$, where $M_i = R_i M$ and $N_i = R_i N$, so that $QM_i \simeq QN_i$ as $QR_i = K_i$-modules. Moreover, M and N are in the same R-genus class iff each M_i and N_i are in the same R_i-genus class. Thus, it suffices to assume that $K = QR$ is a simple algebra and R is a maximal S-order in K. For a prime p, R_p is a maximal S_p-order, M_p, N_p are R_p-lattices with $QM_p \simeq QN_p$, R_p is semi-local, and S_p is a principal ideal domain (Theorem 11.8). Also $S_p\text{-rank}(M_p) = S_p\text{-rank}(N_p)$ since $\dim_F(QM_p) = \dim_F(QN_p)$, where $F = QS$. Since M_p and N_p are free S_p-modules, $\dim_T(M_p/PM_p) = \dim_T(N_p/PN_p)$ where $T = S_p/P\cap S_p$ for each non-zero prime ideal P of R_p. As in the proof of Theorem 11.8, this implies that $M_p \simeq N_p$ as R_p-modules (since M_p and N_p are R_p-projective and $M_p/J(R_p)M_p \simeq N_p/J(R_p)N_p$).

(b) If $p \notin \pi_R$ then $\overline{R}_p = R_p$ so by (a), $M_p \simeq N_p$ if $QM \simeq QN$. ///

Lemma 12.5. Assume that $0 \neq n \in Z$ with $n\overline{R} \subseteq R \subseteq \overline{R}$, X is an R-lattice, and $g : N \to X$ is an R-module epimorphism. If $f \in \mathrm{Hom}_R(X,X)$ then there is $h \in \mathrm{Hom}_R(X,N)$ with $gh = nf$.

Proof. Let $\alpha : F \to X$ be an R-module epimorphism for some free R-module F. Then $f\alpha : F \to X$ so there is $h : F \to N$ with $gh = f\alpha$, since F is projective. Thus, it is sufficient to find $g' : X \to F$ with $\alpha g' = n\cdot 1_X$, in which case $hg' : X \to N$ with $g(hg') = f\alpha g' = nf$, as needed.

Now α lifts to an epimorphism $\alpha' : \overline{R}F \to \overline{R}X$ and $\overline{R}X$ is $\overline{R}$-projective since $\overline{R}X$ is an $\overline{R}$-lattice (Proposition 12.1). Hence there is $g' \in \mathrm{Hom}_{\overline{R}}(\overline{R}X, \overline{R}F)$ with $\alpha'g' = 1_{\overline{R}X}$. Thus $ng' \in \mathrm{Hom}_R(X,F)$ and $\alpha(ng') = \alpha'ng' = n\cdot 1_X$. ///

Theorem 12.6. Suppose that M and N are R-lattices and that M is a genus summand of N. Then $N = N_1 \oplus N_2$ for some R-lattice N_1 in the same genus class as M.

Proof. Choose $n > 1$ in Z with $n\overline{R} \subseteq R \subseteq \overline{R}$. Since M is a genus summand of N there is an integer m relatively prime to n and $f : M \to N$, $g : N \to M$ with $gf = m\cdot 1_M$. Let $X = g(N)$ so that $mM = gf(M) \subseteq g(N) = X \subseteq M$. Write $1 = rm + sn$ for $r, s \in Z$.

Now X and M are R-lattices and if p is a prime divisor of n then $M_p = X_p$, since m and n are relatively prime. Thus, X and M are in the same genus class by Corollary 12.4.b. But $g : N \to X$ is an epimorphism and $s \in \mathrm{Hom}_R(X,X)$ so there is $h \in \mathrm{Hom}_R(X, N)$ with $gh = ns$ (Lemma 12.5). Define $\phi : X \to N$ by $\phi(x) = rf(x) + h(x)$. Then $g\phi(x) = rgf(x) + gh(x) = rmx + snx = x$ for each $x \in X$ so that $N = N_1 \oplus N_2$ with $N_1 \simeq X$. ///

Properties of R-lattices and the notion of genus class and genus summands can be translated into properties of finite rank torsion free groups. If B and C are finite rank torsion free groups then C is a <u>near summand</u> of B if for each $n \in Z$ there is an integer m relatively prime to n and $f : C \to B$, $g : B \to C$ with $gf = m \cdot 1_C$.

<u>Corollary 12.7.</u> Suppose that A is a finite rank torsion free group, $B \in P(A)$ and that $R = E(A)/N(E(A)$. The 1-1 correspondence from isomorphism classes of $P(A)$ to isomorphism classes of $P(R)$ given by $(B) \to (B')$, where $B' = \mathrm{Hom}(A,B)/\mathrm{Hom}(A,B)N(E(A)$, induces 1-1 correspondences from:

(a) isomorphism classes of near summands of B to isomorphism classes of genus summands of B' and

(b) isomorphism classes of finite rank torsion free groups nearly isomorphic to B to isomorphism classes of R-lattices in the same genus class as B'.

<u>Proof.</u> Note that if C is a near summand of B then $C \oplus C' \simeq B \oplus B$ for some C' (as in Lemma 7.20.a). Thus $C \in P(A)$ since $B \oplus B \in P(A)$.

(a) If $f : C \to B$, $g : B \to C$ with $gf = m \cdot 1_C$ then there is $f' : C' \to B'$ and $g' : B' \to C'$ with $g'f' = m \cdot 1_{C'}$ since $B \to B'$ is functorial. Conversely, if $f' : C' \to B'$ and $g' : B' \to C'$ with $g'f' = m$ then there is $f : \mathrm{Hom}(A,C) \to \mathrm{Hom}(A,B)$ and $g : \mathrm{Hom}(A,B) \to \mathrm{Hom}(A,C)$ with $gf(\mathrm{Hom}(A,C)) + \mathrm{Hom}(A,C)N(E(A) = m\mathrm{Hom}(A,C) + \mathrm{Hom}(A,C)N(E(A)$ since $\mathrm{Hom}(A,C)$ and $\mathrm{Hom}(A,B)$ are $E(A)$-projective. Thus $gf(\mathrm{Hom}(A,C)) = m\mathrm{Hom}(A,C)$, by Nakayama's Lemma, so that $m^{-1}gf$ is an automorphism of $\mathrm{Hom}(A,C)$. Hence there is $f_1 = f(gf)^{-1}m : \mathrm{Hom}(A,C) \to \mathrm{Hom}(A,B)$, $g : \mathrm{Hom}(A,B) \to \mathrm{Hom}(A,C)$ with $gf_1 = m \cdot 1$. Now apply Theorem 5.1 to see that C is a near summand of B.

(b) follows from (a). ///

<u>Corollary 12.8.</u> Suppose that A and B are finite rank torsion free groups and let $R = E(A)/N(E(A))$.

(a) If $R = \overline{R}$ then B is nearly isomorphic to A iff B is quasi-isomorphic to A and $B \in P(A)$.

(b) Assume that $R \neq \overline{R}$ and choose a least integer $n > 0$ with $n\overline{R} \subseteq R \subseteq \overline{R}$. Then B is nearly isomorphic to A iff $B \in P(A)$ and there is an integer m relatively prime to n and $f : B \to A$, $g : A \to B$ with $gf = m \cdot 1_B$ and $fg = m \cdot 1_A$.

Proof. (a) ($\Rightarrow$) follows from Lemma 7.20 and the definition of near isomorphism.

($\Leftarrow$) If B is quasi-isomorphic to A then $QB' \simeq QR$ where $B' = \mathrm{Hom}(A,B)/(\mathrm{Hom}(A,B)N(E(A))$. Thus R and B' are in the same genus class by Corollary 12.4.a, so that B is nearly isomorphic to A by Corollary 12.7.

(b) Apply Corollary 12.7 and Corollary 12.4.b. ///

Remark: It is proved in Theorem 13.9, that two finite rank torsion free groups A and B are nearly isomorphic iff there is $0 < k \in Z$ with $A^k \simeq B^k$.

Corollary 12.9. Let A and B be finite rank torsion free groups.

(a) If B is a near summand of A then $A = A_1 \oplus A_2$ for some A_1 nearly isomorphic to B.

(b) If A is nearly isomorphic to $B_1 \oplus B_2$ then $A = A_1 \oplus A_2$ with each A_i nearly isomorphic to B_i.

(c) If A is nearly isomorphic to B then A is indecomposable iff B is indecomposable.

(d) Assume that A is nearly isomorphic to B^k. Then $A \simeq B^{k-1} \oplus C$ for some C nearly isomorphic to B.

Proof. (a) follows from Corollary 12.7 and Theorem 12.6.

(b) If A is nearly isomorphic to $B_1 \oplus B_2$ then B_1 is a near summand of A. Hence $A = A_1 \oplus A_2$ with A_1 nearly isomorphic to B_1 by (a). Thus, A_2 is nearly isomorphic to B_2 by Corollary 7.17.

(c) is a consequence of (b).

(d) By (b), $A = A_1 \oplus \dots \oplus A_k$ with each A_i nearly isomorphic to B. By Lemma 7.20, $A_1 \oplus A_2 \simeq B \oplus B_2$ for some B_2 nearly isomorphic to each A_i. Thus $A = A_1 \oplus A_2 \oplus \dots \oplus A_k \simeq B \oplus B_2 \oplus A_3 \oplus \dots \oplus A_k$. By induction on k, $B_2 \oplus A_3 \oplus \dots \oplus A_k \simeq B^{k-2} \oplus C$ for some C nearly isomorphic to B so that $A \simeq B^{k-1} \oplus C$. ///

Corollary 12.10. Let A be a torsion free group of finite rank.

(a) Every torsion free group nearly isomorphic to A is isomorphic to A iff every projective right ideal of $E(A)/N(E(A))$ in the same genus class as $E(A)/N(E(A))$ is isomorphic to $E(A)/N(E(A))$.

(b) Assume that $E(A)$ is an integral domain. Then every torsion free group nearly isomorphic to A is isomorphic to A iff every invertible ideal of $E(A)$ is principal.

Proof. (a) follows from Corollary 12.7, noting that if B is a subgroup of finite index in A then B is nearly isomorphic to A iff $\mathrm{Hom}(A,B)/\mathrm{Hom}(A,B)N(E(A))$ is a projective ideal and in the same genus class as $E(A)/N(E(A))$.

(b) follows from (a) and the observation that if I is an invertible ideal of $E(A)$ then $I_p \simeq E(A)_p$ for each prime p of Z, since $E(A)_p$ is a semi-local domain so that invertible ideals are principal (as in the proof of Theorem 11.8). ///

Example 12.11. There is a finite rank torsion free group A with a subgroup B of finite index in A such that A is nearly isomorphic to B but not isomorphic to B.

Proof. Let R be any non-principal Dedekind domain such that the additive group of R is finite rank torsion free and reduced (e.g. $R = Z[\sqrt{-5}]$, the ring of algebraic integers in $Q(\sqrt{-5})$). Then $R \simeq E(A)$ for some finite rank torsion free group A (Corner's Theorem, Theorem 2.13). If I is a non-principal projective ideal of $E(A)$, I and $E(A)$ are in the same genus class (Corollary 12.4). Thus, $B = IA \simeq I \otimes_R A$ is a subgroup of finite index in A that is nearly isomorphic to but not isomorphic to A (Corollary 12.10.b).

Example 12.12 (Bass [2]). There is an integral domain S such that $\overline{S}$ is a principal ideal domain, every invertible ideal of S is principal, but S is not a principal ideal domain.

Proof. Let $S = \{a+2bi \mid a,b \in Z\}$ a subring of $\overline{S} = Z[i]$, the ring of Gaussion integers and the ring of algebraic integers in $Q(i)$. Let $I = 2\overline{S} \subsetneq S \subsetneq \overline{S}$. Then

$I = Z2 + Z2i$ is a non-principal ideal of S. Also I is not invertible for if $II^{-1} = S$ then $S = II^{-1} = (\overline{S}I)I^{-1} = \overline{S}S = \overline{S}$, a contradiction.

Let J be an invertible ideal of S. Then $\overline{S}J = \overline{S}f$ for some $f \in \overline{S}$, a principal ideal domain. Thus, $IJ = 2\overline{S}J = 2\overline{S}f = If$ so that $2Jf^{-1}$ is an ideal of S isomorphic to J with $\overline{S}(2Jf^{-1}) = 2\overline{S} = I$ and $I(2Jf^{-1}) = 2I$.

Thus, it suffices to assume that $\overline{S}J = I$ and $IJ = 2I$. Now $J = 2S$, $2Si$ or $Z(2+2i) + 2I$ since $2I \subsetneq J \subsetneq I$ and $I/2I \simeq Z/2Z \oplus Z/2Z$. The latter case is impossible since if $J = Z(2+2i) + 2I$ then $iJ = J$ and $\overline{S}J = J = I$, a contradiction. Thus, J is principal. ///

Example 12.13 (W. J. Leahy). There is an integral domain S such that $\overline{S}$ is a principal ideal domain yet S has an invertible non-principal ideal.

Proof. Let $S = \{a+2b\sqrt{-11} \mid a, b \in Z\}$. Then $\overline{S}$, the ring of algebraic integers in $Q(\sqrt{-11})$ is a principal ideal domain. Let $I = 5S + (1+2\sqrt{-11})S$ a non-principal ideal of S; since if $I = Sr$ then $5 = sr$ for some $s \in S$ while $N(5) = 25 = N(s)N(r)$, where $N(r)$ is the norm of r. Hence s or r is a unit, otherwise $N(s) = a^2 + 44b^2 = 5$; which is impossible. On the other hand $1 + 2\sqrt{-11} \notin S5$ since $N(1+2\sqrt{-11}) = 45$ while $N(t5) = N(t)N(5) = N(t)25$ for each $t \in S$.

Now I is invertible since if $J = S + ((1-2\sqrt{-11})/5)S \subseteq QS$ then $IJ = S5 + S(1+2\sqrt{-11})((1-2\sqrt{-11})/5) + S(1+2\sqrt{-11}) + S(1-2\sqrt{-11}) = S5 + S9 + S(1+2\sqrt{-11}) + S(1-2\sqrt{-11}) = S$. ///

Corollary 12.14. Let $0 < n$ be the least integer with $n\overline{R} \subseteq R \subseteq \overline{R}$ and let I be a left(right) ideal of R with R/I finite.

(a) If $I + nR = R$ then I and R are in the same genus class. Moreover, I is projective and generated by ≤ 2 elements.

(b) If $R = \overline{R}$ then I is a projective R-generator.

Proof. If $n = 1$ then $R = \overline{R}$ and $QI = QR = K$. Thus I and R are in the same genus class by Corollary 12.4.a. Suppose $n > 1$ and let p be a prime divisor of n. Then $R_p = I_p + nR_p = I_p + p^jR_p$ for some j. Thus, $R_p = I_p$ since R_p/I_p is p-divisible and finite. Hence, I and R are in the same genus class by

Corollary 12.4.b. As a consequence of Theorem 12.2, $I \oplus I' \simeq R \oplus R$ for some I' so that I is generated by ≤ 2 elements. This proves (a).

(b) Since $R = \overline{R} = R_1 \times \ldots \times R_k$ and $I = I_1 \times \ldots \times I_k$ with each R_i/I_i finite, then I is an R-generator if each I_i is an R_i-generator. Thus it suffices to assume that $K = QR$ is a simple algebra and that R is a maximal S-order in K. Now apply Theorem 11.3. ///

EXERCISES

12.1 (Lady [3]) Suppose that A and B are finite rank torsion free groups and that A is almost completely decomposable. Prove that the following are equivalent:

(a) A is nearly isomorphic to B;

(b) $A \oplus C \simeq B \oplus C$ for some finite rank completely decomposable group C.

(c) $A \oplus C \simeq B \oplus C$ for some finite rank torsion free group C.

(Hint: (a) $\Rightarrow$ (b), let C be a completely decomposable group of finite index in A and apply Lemma 7.20.)

12.2 Assume that A is almost completely decomposable, say $nA \subseteq A_1^{n_1} \oplus \ldots \oplus A_m^{n_m} \subseteq A$ where each A_i has rank 1 and $A_i \simeq A_j$ iff $i = j$. Prove that if $R = E(A)/N(E(A))$ then $\overline{R} = R_1 \times \ldots \times R_m$ where each R_i is a maximal $E(A_i)$-order in $\mathrm{Mat}_{n_i}(Q)$ and R_i is quasi-equal to the maximal $E(A_i)$-order $\mathrm{Mat}_{n_i}(E(A_i))$. Conclude that $R = \overline{R}$ iff A is completely decomposable. (Hint: Let $B = A_1^{n_1} \oplus \ldots \oplus A_m^{n_m}$, $I = \mathrm{Hom}(A,B)/\mathrm{Hom}(A,B)N(E(A))$ and apply Corollaries 12.14 and 9.6).

§13. Grothendieck Groups

Assume that $\mathcal{C}$ is an additive category such that the isomorphism classes of $\mathcal{C}$ form a set. Define the Krull-Schmidt-Grothendieck group, $K_o(\mathcal{C})$, to be F/R where F is the free abelian group with isomorphism classes of objects of $\mathcal{C}$ as a basis and R is the subgroup of F generated by elements of the form $(A \oplus B) - (A) - (B)$ for each A, B in $\mathcal{C}$. If $A \in \mathcal{C}$ let $[A] = (A) + R \in K_o(\mathcal{C})$. Then $[A \oplus B] = [A] + [B]$ and $n[A] = [A^n]$ if $0 < n \in Z$.

Lemma 13.1. (a) If $x \in K_o(\mathcal{C})$ then $x = [A]-[B]$ for some A, B in $\mathcal{C}$.

(b) $[A] = [B]$ in $K_o(\mathcal{C})$ iff $A \oplus C$ is isomorphic to $B \oplus C$ for some C in $\mathcal{C}$.

Proof. (a) Note that $x = \Sigma\, n_i(A_i) - \Sigma\, m_j(B_j) + R$ for some $0 < n_i \in Z$ and $0 < m_j \in Z$. Thus $x = [\oplus A_i^{n_i}] - [\oplus B_j^{m_j}]$.

(b) ($\Leftarrow$) If $A \oplus C \simeq B \oplus C$ then $(A \oplus C) = (B \oplus C)$ so $[A] + [C] = [B] + [C]$. Since $K_o(\mathcal{C})$ is a group, $[A] = [B]$.

($\Rightarrow$) If $[A] = [B]$ then $(A) - (B) = \Sigma((C_i \oplus D_i) - (C_i) - (D_i)) - (\Sigma_j(E_j \oplus F_j) - (E_j) - (F_j)) \in F$ so that $(A) + \Sigma(C_i) + \Sigma(D_i) + \Sigma(E_j \oplus F_j) = (B) + \Sigma_j(E_j) + \Sigma_j(F_j) + \Sigma_i(C_i \oplus D_i)$. But F is free with isomorphism classes of objects of $\mathcal{C}$ as a basis so $A \oplus (\oplus_i C_i) \oplus (\oplus_i D_i) \oplus (\oplus_j (E_j \oplus F_j)) \simeq B \oplus (\oplus_j E_j) \oplus (\oplus_j F_j) \oplus (\oplus_i (C_i \oplus D_i))$ in $\mathcal{C}$. Therefore, $A \oplus C \simeq B \oplus C$ where $C = (\oplus_i C_i) \oplus (\oplus_i D_i) \oplus (\oplus_j E_j) \oplus (\oplus_j F_j)$. ///

Theorem 13.2. If $\mathcal{C}$ is a Krull-Schmidt category then $K_o(\mathcal{C})$ is a free abelian group with $S = \{[C] \mid C$ is non-zero and indecomposable in $\mathcal{C}\}$ as a basis.

Proof. First of all, S generates $K_o(\mathcal{C})$ for if $x = [A] - [B]$, where $A \simeq \oplus C_i$, $B \simeq \oplus D_j$ with C_i and D_j indecomposable, then

$x = [C_1] + \ldots + [C_m] - [D_1] - \ldots - [D_n]$ is in the subgroup generated by S. Moreover, S is a Z-independent set since if $[C_1], \ldots, [C_k]$ are distinct elements of S with $m_1[C_1] + \ldots + m_k[C_k] = 0$ then $\Sigma\{m_i[C_i] \mid m_i > 0\} = \Sigma\{-m_i[C_i] \mid m_i < 0\}$. Hence if some $m_j \neq 0$ then $(\oplus\{C_i^{m_i} \mid m_i > 0\}) \oplus C \simeq$

$(\oplus C_i^{-mi} | m_i < 0\}) \oplus C$ for some $C \in \mathcal{C}$ by Lemma 13.1.b. Since $\mathcal{C}$ is a Krull-Schmidt category, $\oplus \{C_i^{mi} \mid m_i > 0\} \simeq \oplus \{C_i^{-mi} \mid m_i < 0\}$ so that $C_j \simeq C_k$ for some $j \neq k$, a contradiction. ///

Corollary 13.3. (a) If R is an artinian ring and $\mathcal{M}_R$ the category of finitely generated right R-modules then $K_o(\mathcal{M}_R)$ is a free abelian group.

(b) $K_o(Q\mathcal{A})$ is a free abelian group.

(c) For each prime p, $K_o((\mathcal{A}/p)^*)$ is a free abelian group.

Example 13.4. Let $p \geq 5$ and let $\mathcal{A}_p$ be the category of reduced p-local torsion free abelian groups of finite rank. Then $K_o(\mathcal{A}_p)$ is a free abelian group but $\mathcal{A}_p$ is not a Krull-Schmidt category.

Proof. Example 2.15 shows that $\mathcal{A}_p$ is not a Krull-Schmidt category. The correspondence $(A)_{\mathcal{A}_p} \to (A)_{(\mathcal{A}/p)^*}$ induces a homomorphism $\phi : K_o(\mathcal{A}_p) \to K_o((\mathcal{A}/p)^*)$ with $\phi([A]-[B]) = [A]-[B]$. Now Kernel$(\phi) = \{[A]-[B] \mid [A] = [B] \in K_o((\mathcal{A}/p)^*)\}$. If $[A] = [B] \in K_o((\mathcal{A}/p)^*)$ then $A \oplus C \simeq B \oplus C$ in $(\mathcal{A}/p)^*$ for some C in $(\mathcal{A}/p)^*$. Thus $A \simeq B$ in $(\mathcal{A}/p)^*$, a Krull-Schmidt category, so that $A \simeq B$ in $P\mathcal{A}$ (Theorem 7.11 and Theorem 7.13) hence in $\mathcal{A}_p$. Consequently, ϕ is a monomorphism so that $K_o(\mathcal{A}_p)$ is a free abelian group since $K_o((\mathcal{A}/p)^*)$ is free. ///

Lemma 13.5. Let $P = \Pi Z$ be a product of $|I|$ copies of Z and let $G = \{(n_i) \in P \mid$ for some $0 \leq n \in Z$, $|n_i| \leq n$ for each $i \in I\}$. Then G is a free abelian group.

Proof. Note that G is a pure subgroup of P and G is a commutative ring. Furthermore, G is generated as a ring by the set $E = \{x_J | J \subseteq I\}$ of idempotents of J where $x_J = (n_i)$, $n_i = 0$ if $i \notin J$ and $n_i = 1$ if $i \in J$.

Thus it is sufficient to prove a theorem of G.M. Bergman: If R is a commutative ring with 1, R^+ is torsion free, and R is generated as a ring by a set E of idempotents then R^+ is a free abelian group. (The following proof is due to R.S. Pierce).

The proof is by induction on the cardinality of E. If $E = \phi$ then $R = Z\cdot 1_R$. Let $E = \{e_\sigma | \sigma < \tau\}$ where τ is the smallest ordinal with card τ = card E.

For $\rho \leq \tau$ define R_ρ to be the subring of R generated by $E_\rho = \{e_\sigma | \sigma < \rho\}$. Then $R_1 = Z\cdot 1_R$, $R_\tau = R$, $R_\sigma \subseteq R_{\sigma+1}$ and $R_\rho = \cup\{R_\sigma | \sigma < \rho\}$ if ρ is a limit ordinal $\leq \tau$.

Assume, for the moment, that R_σ is pure in R for each $\sigma < \tau$. Then $\overline{R} = e_\sigma R_\sigma / R_\sigma \cap e_\sigma R_\sigma$ is a commutative ring generated by a set of idempotents $\overline{E} = \{\overline{e_\sigma e_\rho} | \rho < \sigma\}$, since $e_\sigma R_\sigma$ is a commutative ring with identity e_σ generated by idempotents $\{e_\sigma e_\rho | \rho < \sigma\}$. Also $\overline{R}^+$ is torsion free, since as groups $e_\sigma R_\sigma / R_\sigma \cap e_\sigma R_\sigma \simeq (R_\sigma + e_\sigma R_\sigma)/R_\sigma = R_{\sigma+1}/R_\sigma$ which is torsion free, noting that $R_{\sigma+1}/R_\sigma \subseteq R/R_\sigma$ which is torsion free by assumption. By induction on $|E|$, $\overline{R}^+ \simeq R_{\sigma+1}/R_\sigma$ is a free abelian group. Thus $R^+ \simeq \Sigma_{\sigma<\tau} \oplus (R_{\sigma+1}/R_\sigma)^+$ is free as desired.

To prove that R_σ is pure in R, let F be a finite subset of E. Then $1 = \Pi_{e\in F}((1-e)+e) = \Sigma_{S\subseteq F} f_S$ where $f_S = (\Pi_{e\in S} e)(\Pi_{e\in F\setminus S}(1-e))$. Then $\{f_S | S \subseteq F\} = \{f_1, \ldots, f_n\}$ is an orthogonal set of idempotents with $1 = f_1 + \ldots + f_n$. Hence $\langle F\rangle$, the subring of R generated by F, is $Zf_1 \times \ldots \times Zf_n$ since if $e \in F$ then $e = \Sigma\{f_j | ef_j \neq 0\} = \Sigma\{f_S | e \in S\} = \Sigma\{f_j | ef_j = f_j\}$.

Now let $r \in R$, $s \in R_\sigma$, $0 \neq n \in Z$, $nr = s$. Then $s \in \langle F_1\rangle = Zg_1 \times \ldots \times Zg_\ell \subseteq \langle F_2\rangle = Zf_1 \times \ldots \times Zf_k$ where $r \in \langle F_2\rangle$, $\{g_i\} \subseteq F_2$, $F_1 \subseteq F_2$ are finite subsets of E, $F_1 \subseteq E_\sigma$, and $\{g_i\}$, $\{f_j\}$ are orthogonal sets of idempotents with $1 = g_1 + \ldots + g_\ell = f_1 + \ldots + f_k$. Let $r = \sum_{j=1}^{k} a_j f_j$, $s = \sum_{i=1}^{\ell} b_i g_i$. Since $nr = s$, $\sum_{j=1}^{k} na_j f_j = \sum_{i=1}^{\ell} b_i g_i$. As above, each $g_i = \Sigma\{f_j | g_i f_j = f_j\}$. Given $1 \leq i' \leq \ell$, and j' with $g_{i'} f_{j'} = f_{j'}$ then $na_{j'} f_{j'} = f_{j'}(\Sigma na_j f_j) = \sum_{i=1}^{\ell} f_{j'}(b_i g_i) = b_{i'} f_{j'} g_{i'}$ (g_i and $g_{i'}$ are orthogonal if $i \neq i'$ so $g_{i'} f_{j'} = f_{j'}$ implies that $g_i f_{j'} = 0$) $= b_{i'} f_{j'}$. Since R is torsion free $na_{j'} = b_{i'} \in Z$. Hence $r = (1/n)s = \Sigma(b_i/n)g_i \in R_\sigma$ as desired. ///

Theorem 13.6 (Lady [2]). Let $\bar{A}$ be the category of reduced torsion free abelian groups of finite rank.

(a) $T = \{[A] - [B] \mid A$ is nearly isomorphic to $B\}$ is the torsion subgroup of $K_o(\bar{A})$.

(b) $K_o(\bar{A})/T$ is a free abelian group.

Proof. The correspondence $[A] \to ([A]_{(A/p)^*})$ induces a homomorphism $\phi : K_o(\bar{A}) \to \Pi_p K_o((A/p)^*)$. Then $\mathrm{Kernel}(\phi) = \{[A] - [B] \mid$ for each p there is C with $A \oplus C \simeq B \oplus C$ in $(A/p)^*\} = T$ (Theorem 7.11 and Theorem 7.13).

In fact, Image(ϕ) is a free abelian group. By Corollary 13.3 $K_o((A/p)^*) = \oplus Zx_{p,\alpha}$ where $x_{p,\alpha} = [C_{p,\alpha}]$, $C_{p,\alpha}$ indecomposable in $(A/p)^*$. Thus Image$(\phi) \subseteq \Pi_p K_o((A/p)^*) \subseteq \Pi_p \Pi_\alpha Zx_{p,\alpha} = P$. But $\phi([A]) = ([A]_{(A/p)^*}) = (\Sigma n_{p,\alpha}x_{p,\alpha}) \in P$, where for each p, almost all $n_{p,\alpha} = 0$ and $0 \leq n_{p,\alpha} \leq (\mathrm{rank}(A))^2$ since A is a direct sum of finitely many indecomposables in $(A/p)^*$ obtained from idempotents of $E(A)/pE(A)$. Hence Image$(\phi) \subseteq \{(n_{p,\alpha}x_{p,\alpha}) \mid$ for some $n \in Z$, $|n_{p,\alpha}| \leq n$ for each p, $\alpha\} = G$. By Lemma 13.5, G is a free abelian group, whence Image(ϕ) is a free abelian group.

Since Image$(\phi) = K_o(\bar{A})/T$ is free, the torsion subgroup of $K_o(\bar{A})$ is contained in T. Thus, it suffices to prove that if $[A] - [B] \in K_o(\bar{A})$ with A nearly isomorphic to B then $n([A]-[B]) = 0$ for some $0 \neq n \in Z$.

Fix A and let $U_A = \{[B]-[A] \mid B$ is nearly isomorphic to $A\}$. Then U_A is a subgroup of $K_o(\bar{A})$ for if B and C are nearly isomorphic to A then $A \oplus D \simeq B \oplus C$ for some D nearly isomorphic to A (Lemma 7.20). Thus $[B]-[A] + [C]-[A] = [D]-[A] \in K_o(\bar{A})$. Also, if $[B]-[A] \in U_A$ and $B \oplus C \simeq A \oplus A$ for some C nearly isomorphic to A then $[B] - [A] + [C] - [A] = 0$, i.e. $[C] - [A] = -([B]-[A]) \in U_A$. Finally, if $[B] - [A] \in U_A$ then B is isomorphic to a summand of $A \oplus A$ so that U_A is finite as a consequence of Theorem 11.11. Hence $n([B]-[A]) = 0 \in K_o(\bar{A})$ for some $0 < n \in Z$. ///

Corollary 13.7. Assume that A and B are torsion free groups of finite rank. Then A and B are nearly isomorphic iff there is a finite rank torsion free group C and $0 < n \in Z$ with $A^n \oplus C \simeq B^n \oplus C$.

Proof. ($\Leftarrow$) is Corollary 7.17.

($\Rightarrow$) It is sufficient to assume that A and B are reduced. By Theorem 13.6, $n([A]-[B]) = 0$ for some $0 < n \in Z$ so that $[A^n] = [B^n]$ and $A^n \oplus C \simeq B^n \oplus C$ for some C (Lemma 13.1). ///

Let C be an additive category and A an object of C. Define $\underline{K_o(A)} = K_o(P(A))$ where $P(A)$ is the category of summands of finite direct sums of copies of A. For example, if $C = M_R$ then $K_o(R) = K_o(P(R))$ is the usual Grothendieck group of finitely generated projective R-modules.

Corollary 13.8. Let A be a finite rank torsion free group.

(a) $K_o(A) \simeq K_o(\overline{E(A)})$, where $\overline{E(A)} = E(A)/N(E(A)$

(b) The torsion subgroup of $K_o(A)$ is $T_A = \{[B] - [A] \mid$ B is nearly isomorphic to A$\}$.

(c) The torsion subgroup of $K_o(\overline{E(A)})$ is $T_{E(A)} = \{[M] - [\overline{E(A)}]$ and M, $\overline{E(A)}$ are in the same genus class$\}$.

(d) $T_A \simeq T_{\overline{E(A)}}$.

Proof. (a) is a consequence of Corollary 9.6.

(b) As in Theorem 13.6, $T_A = \{[B]-[C] \mid B, C \in P(A)$ and B nearly isomorphic to C$\}$. If $[B] - [C] \in T_A$ then $C \oplus C' \simeq A^n$ for some $0 < n \in Z$ and some C'. Thus $[B]-[C] = [B\oplus C'] - [C\oplus C'] = [B\oplus C'] - [A^n]$ with $B \oplus C'$ nearly isomorphic to $C \oplus C' \simeq A^n$ and $T_A = \{[B] - [A^n] \mid 0 < n \in Z,$ B nearly isomorphic to $A^n\}$. But if B is nearly isomorphic to A^n then $B \simeq A^{n-1} \oplus B'$ for some B' nearly isomorphic to A (Corollary 12.9.d). Hence $[B]-[A^n] = [A^{n-1}] + [B'] - [A^n] = (n-1)[A] + [B'] - n[A] = [B'] - [A]$ with B' nearly isomorphic to A. This proves (b).

(c) and (d) are consequences of (b) and Corollary 12.7. ///

The following theorem is a rather striking characterization of near-isomorphism, due originally to R.B. Warfield, Jr., as a consequence of the fact that 2 is in the stable range of E(A) for a finite rank torsion free group A.

Theorem 13.9. Let A and B be torsion free groups of finite rank. Then A and B are nearly isomorphic iff there is $0 < k \in Z$ with $A^k \simeq B^k$.

Proof. ($\Leftarrow$) is Corollary 7.17.b.

($\Rightarrow$) There is $0 < k \in Z$ with $k([B]-[A]) = 0 \in K_o(A)$, by Corollary 13.8, so that $B^k \oplus C \simeq A^k \oplus C$ for some C in $P(A)$. But $C \oplus C' \simeq A^n$ for some $0 < n \in Z$ so that $B^k \oplus A^n \simeq A^{k+n}$.

Assume that $n \geq 2$. Then $B^k \oplus A^{n-1} \simeq A^{n+k-2} \oplus A'$ for some $A \oplus A' \simeq A \oplus A$ (Theorem 8.15.b) so that $B^k \oplus A^{n-1} \simeq A^{n+k-1}$. Hence, it suffices to assume that $B^k \oplus A \simeq A^{k+1}$, i.e. $n = 1$.

Assume that $k \geq 2$. Then $B^k \simeq A' \oplus A^{k-1}$, where $A' \oplus A \simeq A \oplus A$, by Theorem 8.15.b, so that $B^k \simeq A^k$.

The only case left is $B \oplus A \simeq A \oplus A$. Since B is nearly isomorphic to A, $A \oplus A' \simeq B \oplus B$ for some A' nearly isomorphic to A (Lemma 7.20). Thus $B \oplus B \oplus B \simeq B \oplus A \oplus A' \simeq A \oplus A \oplus A' \simeq A \oplus B \oplus B \simeq A \oplus A \oplus B$ so that $A \oplus A \simeq B \oplus B'$ for some $B \oplus B' \simeq B \oplus B$. Hence, $A \oplus A \simeq B \oplus B$. ///

Corollary 13.10. Let A be a torsion free group of finite rank. Then $K_o(A)$ is a finitely generated abelian group, T_A is bounded by k = the number of distinct isomorphism classes of summands of $A \oplus A$, and $K_o(A)/T_A$ is a free abelian group. If $\overline{E(A)/N(E(A))} = E(A)/N(E(A))$ (notation as in Section 12) then $K_o(A)/T_A \simeq Z$.

Proof. Note that $T_A = \{[B]-[A] \mid B$ is nearly isomorphic to $A\}$ is finite and $kT_A = 0$ since if B is nearly isomorphic to A then $B \oplus B' \simeq A \oplus A$ for some B' and $A \oplus A$ has, up to isomorphism, only k summands.

Let $R = E(A)/N(E(A))$ so that $K_o(R) \simeq K_o(A)$ and $T_R \simeq T_A$, where $T_R = \{[M]-[R] \mid M$ and R are in the same genus class$\}$ (Corollary 13.8).

Define $\phi : K_o(R) \to Z \oplus (\oplus\{K_o(R/pR) \mid p \text{ divides } n\})$, where $0 < n$ is the least integer with $n\overline{R} \subseteq R \subseteq \overline{R}$ as in Section 12, by $\phi(M) = (\text{rank}(M), [M/p_1M], \ldots, [M/p_mM])$ where $\{p_1, \ldots, p_m\}$ is the set of prime divisors of n. If $[M]-[N] \in \text{Kernel}(\phi)$ then $QM \simeq QN$ and $[M/p_iM] = [N/p_iN] \in K_o(R/p_iR)$ for each i. Since R/p_iR is finite, hence artinian, $M/p_iM \simeq N/p_iN$ for each i. But M and N are R-projective so that M_{p_i} and N_{p_i} are R_{p_i}-projective with $M_{p_i}/p_iM_{p_i} \simeq N_{p_i}/p_iN_{p_i}$. Thus

$M_{p_i} \simeq N_{p_i}$ as R_{p_i}-modules for each i since $p_iR_{p_i} \subseteq J(R_{p_i})$, whence M and N are in the same genus class (Corollary 12.4). Therefore, $[M]-[N] \in T_R$ and $\text{Kernel}(\phi) \subseteq T_R$. But $\text{Image}(\phi)$ is a finitely generated free group so that $T_R \subseteq \text{Kernel}(\phi)$ and $K_o(R)/T_R = \text{Image}(\phi)$ is a finitely generated free group, noting that each $K_o(R/p_iR)$ is a finitely generated free group. ///

Suppose that S is an integral domain, $S \subseteq QS$, so that QS is an algebraic number field. Define the ideal class group, $I(S)$, to be the set of isomorphism classes of invertible ideals of S.

Lemma 13.12. (a) Two ideals I and J of S are S-isomorphic iff there are non-zero elements x and y of S with $xI = yJ$.

(b) $I(S)$ is an abelian group, where $(I)(J) = (IJ)$, and (S) is the identity.

(c) If $I_1, \ldots, I_n$ and $J_1, \ldots, J_m$ are ideals of $E(A)$ with $I_1 \oplus \ldots \oplus I_n \simeq J_1 \oplus \ldots \oplus J_m$ then $m = n$ and $I_1I_2 \ldots I_n$ is S-isomorphic to $J_1J_2 \ldots J_n$.

Proof. (a) ($\Leftarrow$) Since S is a domain, $I \simeq xI = yJ \simeq J$.

($\Rightarrow$) If $\phi : I \to J$ is an isomorphism and $0 \neq y \in I$, $x = \phi(y)$ then $xI = yJ$ since if $r \in I$ then $y\phi(r) = \phi(yr) = \phi(y)r = xr$.

(b) The operation is well defined, associative, and (S) is the identity. If $(I) \in I(S)$ then $II^{-1} = S$. Since I^{-1} is a finitely generated S-submodule of QS, $J = nI^{-1}$ is an invertible ideal for some $0 \neq n \in Z$ and $(I)(J) = (nS) = (S)$.

(c) Let $\phi : I_1 \oplus \ldots \oplus I_n \to J_1 \oplus \ldots \oplus J_m$ be an isomorphism and let $\phi_{ij} : I_i \to J_j$ be ϕ restricted to I_i followed by a projection to J_j. By (a) $\phi_{ij}(x_i) = q_{ij}x_i$ for some $q_{ij} \in QS$ and each $x_i \in I_{ij}$. If $M = (q_{ij})$ an $m \times n$ matrix with entries in QS then M is invertible so $m = n$. But $(\det M)I_1I_2 \ldots I_n \subseteq J_1J_2 \ldots J_m$, since each $q_{ij}x_i = \phi_{ij}(x_i) \in J_j$. Similarly, $\det(M^{-1})J_1J_2 \ldots J_m \subseteq I_1I_2 \ldots I_n$. Thus $J_1J_2 \ldots J_n = (\det M)I_1I_2 \ldots I_n$. Now apply (a).

Theorem 13.13. Suppose that A is finite rank torsion free and that $S = E(A)/N(E(A))$ is an integral domain. Then T_A, the torsion subgroup of $K_o(A)$, is isomorphic to $I(S)$ and $I(S)$ is finite.

Proof. Since $K_o(A) \simeq K_o(S)$ and $T_S \simeq T_A$, it suffices to prove that $T_S \simeq I(S)$.

Define $\phi : T_S \to I(S)$ by $\phi([I]-[S]) = (I)$, noting that if $x \in T_S$ then $x = [I]-[S]$ for some S-lattice I in the same genus class as S so that I is isomorphic to an ideal of S. If I is an invertible ideal of S then $I_p \simeq S_p$ for each p (Lemma 11.8) so that I and S are in the same genus class. Hence ϕ is onto also ϕ is 1-1 since if $(I) = (S)$ then $I \simeq S$ so that $[I] = [S] \in K_o(S)$.

It remains to prove that ϕ is a well defined group homomorphism. If $[I]-[S] = [J]-[S]$ then $[I] = [J]$ so $I \oplus S^n \simeq J \oplus S^n$ for some n. By Lemma 13.12.c, $(I) = (J)$ proving that ϕ is well defined. If $[I] - [S]$ and $[J] - [S] \in T_S$ then $[I]-[S] + [J] - [S] = [K] - [S]$ where $S \oplus K \simeq I \oplus J$. Again by Lemma 13.12.c, $(K) = (SK) = (IJ) = (I)(J)$ so that ϕ is a group homomorphism. ///

Corollary 13.14. Suppose that A is a torsion free group, $R = E(A)/N(E(A))$ and that $R = \overline{R}$. Then $K_o(A)$ is isomorphic to a subgroup of $K_o(\mathcal{A})$. Moreover, if S is the ring of algebraic integers of an algebraic number field then $I(S)$ is isomorphic to a subgroup of the torsion subgroup of $K_o(\mathcal{A})$.

Proof. Define $\phi : K_o(A) \to K_o(\mathcal{A})$ by $\phi([B]-[C]) = [B] - [C]$. If $x = [D] - [C] \in \mathrm{Kernel}(\phi)$ then $x = [B] - [A^n]$ where $C \oplus C' \simeq A^n$ and $B \oplus G \simeq A^n \oplus G$ for some G in $\mathcal{A}$.

It suffices to prove that $B \oplus C \simeq A^n \oplus C$ for some C in $P(A)$. Let $G' = \mathrm{Hom}(A,G)/\mathrm{Hom}(A,G)N(E(A))$ and write $G' = M \oplus N$, where M is a projective R-module and N has no projective R-summands. Then $B' \oplus G' = B' \oplus M \oplus N \simeq R^n \oplus M \oplus N = R^n \oplus G'$. But $\mathrm{Hom}_R(N, R^n \oplus M) = \mathrm{Hom}_R(N, B' \oplus M) = 0$ since $R^n \oplus M$ and $B' \oplus M$ are R-projective, R is right hereditary (since $R = \overline{R}$, Corollary 11.5) and N has no projective summands. Thus, $B' \oplus M \simeq R^n \oplus M$ since under the isomorphism $B' \oplus M \oplus N \to R^n \oplus M \oplus N$, N is mapped isomorphically to N. Consequently, $B \oplus X \simeq A^n \oplus X$ for some $X \in P(A)$ (Corollary 9.6) and $x = [B] - [A^n] = 0 \in K_o(A)$.

If S is the ring of algebraic integers of an algebraic number field then S is finitely generated and free as an abelian group so $E(A) \simeq S$ for some finite rank torsion free group A (Corner's Theorem). Moreover S is algebraically closed so $S = \overline{S}$. Also $\mathcal{I}(S) \simeq T_A$ is finite (Theorem 13.13) so that $\mathcal{J}(S)$ is isomorphic to a subgroup of the torsion subgroup of $K_0(\mathcal{A})$. ///

Let $\mathcal{C}$ be an additive category of modules and define the Jordan-Hölder-Grothendieck group, $K^0(\mathcal{C})$, to be $\mathcal{F}/\mathcal{R}$ where $\mathcal{F}$ is the free abelian group with isomorphism classes of objects of $\mathcal{C}$ as a basis and $\mathcal{R}$ is the subgroup generated by $(B)-(A)-(C)$ such that A, B, C are in $\mathcal{C}$ and there is an exact sequence of modules $0 \to A \to B \to C \to 0$. For A in $\mathcal{C}$ let $[A] = (A) + \mathcal{R}$. Hence if $0 \to A \to B \to C \to 0$ is an exact sequence of modules then $[B] = [A] + [C]$.

Proposition 13.15. (a) If $A \simeq B$ then $[A] = [B] \in K^0(\mathcal{C})$.

(b) $[A \oplus B] = [A] + [B]$.

(c) There is an epimorphism $K_0(\mathcal{C}) \to K^0(\mathcal{C})$.

(d) If each exact sequence of modules in $\mathcal{C}$ is split exact then $K_0(\mathcal{C}) \simeq K^0(\mathcal{C})$.

(e) $K_0(\mathcal{P}(R)) \simeq K^0(\mathcal{P}(R))$ for a ring R.

Proof. (a) and (b) are clear.

(c) Define $\phi : K_0(\mathcal{C}) \to K^0(\mathcal{C})$ by $\phi([A]-[B]) = [A]-[B]$, a well defined homomorphism by (a) and (b). Clearly, ϕ is an epimorphism.

(d) In this case $K_0(\mathcal{C})$ and $K^0(\mathcal{C})$ have the same generators and the same relations.

(e) follows from (d). ///

Lemma 13.16. (a) If $x \in K^0(\mathcal{C})$ then $x = [A]-[B]$ for some A, B in $\mathcal{C}$.

(b) $[A] = [B] \in K^0(\mathcal{C})$ iff there are exact sequences $0 \to C' \to C \to C'' \to 0$ and $0 \to D' \to D \to D'' \to 0$ of objects of $\mathcal{C}$ such that $A \oplus C' \oplus C'' \oplus D \simeq B \oplus C \oplus D' \oplus D''$.

Proof. Imitate the proof of Lemma 13.1. ///

Theorem 13.17 (Rotman [2]). Let A be the category of finite rank torsion free groups.

(a) $[A] = [B] \in K^o(A)$ iff rank(A) = rank(B) and p-rank(A) = p-rank(B) for each prime p of Z.

(b) $K^o(A)$ is isomorphic to Z⊕G where G is the group of bounded sequences in a countable product of copies of Z.

(c) There is an exact sequence $0 \to G \to K_o(A) \to K^o(A) \to 0$ where $G = \{[A]-[B] \mid \text{rank}(A) = \text{rank}(B)$ and p-rank(A) = p-rank(B) for each prime p of Z$\}$.

Proof. (a) (⟹) If $[A] = [B]$ then $A \oplus C' \oplus C'' \oplus D \simeq B \oplus C \oplus D' \oplus D''$ where $0 \to C' \to C \to C'' \to 0$ and $0 \to D' \to D \to D'' \to 0$ are exact sequences. Thus p-rank(C') + p-rank(C'') = p-rank(C) and p-rank(D') + p-rank(D'') = p-rank(D) from which it follows that p-rank(A) = p-rank(B). Similarly, rank(A) = rank(B).

(⟸) There is a chain $0 = A_0 \subseteq A_1 \subseteq \dots \subseteq A_n = B$ such that n = rank(B) and each A_{i+1}/A_i is rank-1 torsion free. Thus $[B] = [A_n/A_{n-1}] + \dots + [A_1/A_0]$. If A is a rank-1 group then there is a rank-2 torsion free group G with exact sequences $0 \to A \to G \to Q \to 0$ and $0 \to E(A) \to G \to Q \to 0$ (Example 2.6). Hence $[A] = [G]-[Q] = [E(A)]$. Thus $[A_n/A_{n-1}] + \dots + [A_1/A_0] = [E(A_n/A_{n-1})] + \dots + [E(A_1/A_0)]$ so that $[B] = [C_1 \oplus \dots \oplus C_n]$ where each C_i is a rank-1 group with non-nil type, p-rank(B) = p-rank(C_1) + ... + p-rank(C_n), and rank(B) = rank(C_1) + ... + rank(C_n) = n.

If X and Y are rank-1 groups then there is an exact sequence $0 \to X_1 \to X \oplus Y \to Y_1 \to 0$ where X_1 and Y_1 are rank-1 groups with type(X_1) = IT(X⊕Y) and type(Y_1) = OT(X⊕Y). Thus $[X \oplus Y] = [X_1] + [Y_1]$ with type(X_1) ≤ type(Y_1) and p-rank (X ⊕ Y) = p-rank(X_1) + p-rank(Y_1).

It now suffices to prove that if $A = D_1 \oplus \dots \oplus D_n$, $B = E_1 \oplus \dots \oplus E_n$ where each D_i and E_j are rank-1 groups with non-nil type, type(D_1) ≤ ... ≤ type(D_n), type(E_1) ≤ ... ≤ type(E_n), and p-rank(A) = p-rank(B) for each prime p then $A \simeq B$.

If p-rank(A) = k then $pD_i \neq D_i$ for $1 \le i \le k$ and $pD_i = D_i$ if $k + 1 \le i \le n$. Thus type(D_n) = $[(m_p)]$ where $m_p = 0$ if p-rank(A) ≥ n and

$m_p = \infty$ if p-rank(A) < n. Hence $\text{type}(E_n) = \text{type}(D_n)$ so that $D_n \simeq E_n$. By induction on n, $A \simeq B$ and [A] = [B].

(b) Define $\phi : K^o(\mathcal{A}) \to \pi_p Z$ by $\phi([A]-[B]) = (\text{p-rank}(A) - \text{p-rank}(B))$, a well defined homomorphism (0 is included as a prime where 0-rank(A) = rank(A)). If $x = [A] - [B] \in \text{Kernel}(\phi)$ then p-rank(A) = p-rank(B) for each p so that $x = 0 \in K^o(\mathcal{A})$ by (a). Thus ϕ is a monomorphism. But p-rank(A) $\leq$ rank(A) for each prime p so that Image(ϕ) $\subseteq$ bounded sequences of $\pi_p Z$, where Image(ϕ) is a free group by Lemma 13.5. In fact, Image(ϕ) $\simeq$ Z$\oplus$G where G is the group of bounded sequences in $\pi_{p\neq 0} Z$ (Exercise 13.3).

(c) is a consequence of (a) and Proposition 13.15.c. ///

EXERCISES

13.1 Let A be a finite rank torsion free group such that E(A) is an integral domain.

(a) Prove that $K_0(E(A))$ is a commutative ring with identity $[E(A)]$ where $[M][N] = [M \otimes_{E(A)} N]$.

(b) Use the isomorphism $K_0(E(A)) \to K_0(A)$ to induce a ring structure on $K_0(A)$ and identify multiplication in $K_0(A)$.

(c) Prove that if T is the torsion subgroup of $K_0(E(A))$ then T is an ideal of $K_0(E(A))$ and $T^2 = 0$.

13.2 For A, B in $\mathcal{A}$ define $A \equiv B$ if $A \in \mathcal{P}(B)$ and $B \in \mathcal{P}(A)$.

(a) Show that $\equiv$ is an equivalence relation on the set of isomorphism classes of objects of $\mathcal{A}$.

(b) Partially order the set of equivalence classes by $[A] \leq [B]$ if $A \in \mathcal{P}(B)$ and observe that $[A] = [B]$ iff $\mathcal{P}(A) = \mathcal{P}(B)$.

(c) For $[A] \leq [B]$ define $\phi_A^B : K_0(A) \to K_0(B)$ by $\phi_A^B([C]) = [C]$.

(d) Prove that $K_0(\mathcal{A}) \simeq \varinjlim \{K_0(A), \phi_A^B\}$.

13.3 Prove that $K^0(\mathcal{A}) \simeq Z \oplus G$ where G is the set of bounded sequences in $\pi_{p \neq 0} Z$.

13.4 Show that the Jordan-Hölder Theorem fails in $\mathcal{A}$ by exhibiting a rank-2 torsion free group A with pure rank-1 subgroups A_1 and A_2 such that $A_2 \not\simeq A_1$ and $A_2 \not\simeq A/A_1$.

§14. Additive Groups of Subrings of Finite Dimensional Q-Algebras

The notation for this section is:

K - finite dimensional Q-algebra,

R - subring of K with $QR = K$,

R+ - additive group of R,

F - Center(K),

S - Center(R) so that $QS = F$.

Theorem 14.1 (Wedderburn Principal Theorem): There is a subalgebra K' of K with $K = K' \oplus \mathcal{J}(K)$ as vector spaces.

Proof. Let $J = \mathcal{J}(K)$. If $J = 0$ there is nothing to prove. Now assume that $J \neq 0$ but $J^{(2)} = 0$. Write $K/J = K_1 \times \ldots \times K_k$ as a product of simple algebras so that $C(K/J) = F_1 \times \ldots \times F_k$, where each $F_i = C(K_i)$ is an algebraic number field. By Lemma 10.7 there is an algebraic number field $E_i \supseteq F_i$ such that $E_i \otimes_{F_i} K_i \simeq \mathrm{Mat}_{n_i}(E_i)$ where $n_i^2 = \dim_{F_i}(K_i)$ for each i. Write $F_i = Q[X]/\langle f_i(X)\rangle$ where $f_i(X) \in Q[X]$ is a separable polynomial. Let E_i' be a splitting field of f_i over Q. There is an algebraic number field E containing copies of E_i and E_i' for each i. Thus $E \otimes_Q F_i \simeq E[X]/\langle f_i(X)\rangle \simeq \pi E$. For each i, $E \otimes_Q K_i \simeq (E \otimes_Q F_i) \otimes_{F_i} K_i \simeq \pi(E \otimes_{F_i} K_i)$ and $E \otimes_{F_i} K_i \simeq E \otimes_{E_i} (E_i \otimes_{F_i} K_i) \simeq E \otimes_{E_i} \mathrm{Mat}_{n_i}(E_i) \simeq \mathrm{Mat}_{n_i}(E \otimes_{E_i} E_i) \simeq \mathrm{Mat}_{n_i}(E)$. Therefore, $E \otimes_Q (K/J) \simeq \pi(E \otimes_Q K_i) \simeq \pi \mathrm{Mat}_{n_i}(E)$ is a product of matrix rings over the field E.

Now $E \otimes_Q (K/J) \simeq (E \otimes_Q K)/\mathcal{J}(E \otimes_Q K)$: noting that $0 \to E \otimes_Q J \to E \otimes_Q K \to E \otimes_Q (K/J) \to 0$ is exact; $E \otimes_Q J$, being a nilpotent ideal of $E \otimes_Q K$, is contained in $\mathcal{J}(E \otimes_Q K)$; and thus $E \otimes_Q J = \mathcal{J}(E \otimes_Q K)$ since $(E \otimes_Q K)/(E \otimes_Q J) \simeq E \otimes_Q (K/J) \simeq \pi \mathrm{Mat}_{n_i}(E)$ is a semi-simple algebra and $\mathcal{J}(E \otimes_Q K)/E \otimes_Q J)$ is a nilpotent ideal of $(E \otimes_Q K)/(E \otimes_Q J)$.

Therefore, $(E \otimes_Q K)/\mathcal{J}(E \otimes_Q K) \simeq \pi \mathrm{Mat}_{n_i}(E)$ is a product of matrix rings over E. By Exercise 14.1, $E \otimes_Q K$ contains a subalgebra $L \simeq (E \otimes_Q K)/\mathcal{J}(E \otimes_Q K)$. Since L is semi simple, $L \cap \mathcal{J}(E \otimes_Q K) = 0$ (being a nilpotent ideal of L). Consequently, $E \otimes_Q K = L \oplus \mathcal{J}(E \otimes_Q K)$ as vector spaces since $\dim_Q(E \otimes_Q K) = \dim_Q(L) + \dim_Q(\mathcal{J}(E \otimes_Q K))$.

The next step is to prove that if $J^2 = 0$ with $J \neq 0$ then $K = K' \oplus J$ for some subalgebra K' of K. Let $\{u_1 + J, \ldots, u_m + J\}$ be a Q-basis of K/J. Since $K \subseteq E\otimes_Q K = L \oplus \mathcal{J}(E\otimes_Q K)$, each $u_i = u_i' + w_i$ for $u_i' \in L$, $w_i \in \mathcal{J}(E\otimes_Q K)$. Then $\{u_1', \ldots u_m'\}$ is an E-basis of L since $\{u_i' + \mathcal{J}(E\otimes_Q K)\} = \{u_i + \mathcal{J}(E\otimes_Q K)\}$ and $(E\otimes_Q K)/\mathcal{J}(E\otimes_Q K) \simeq E\otimes_Q(K/J) \simeq L$. In fact, $Qu_1' \oplus \ldots \oplus Qu_m'$ is a subalgebra of L, since $(u_i+J)(u_j+J) = \Sigma_k q_{ijk}(u_k+J)$ with $q_{ijk} \in Q$ implies that $u_i' u_j' - \Sigma_k q_{ijk} u_k' \in \mathcal{J}(E\otimes_Q K) \cap L = 0$ and $u_i' u_j' = \Sigma_k q_{ijk} u_k'$. Let $\{e_1 = 1, e_2, \ldots, e_r\}$ be a Q-basis of E. Then each $w_i \in \mathcal{J}(E\otimes_Q K) = E\otimes_Q J$, say $w_i = \Sigma_j e_j w_{ij}$ with $w_{ij} \in J$. Define $y_i = u_i - w_{i1} \in K$ so that $K' = Qy_1 \oplus \ldots \oplus Qy_m$ is a subspace of K. Moreover $K' \cap J = 0$, by the choice of $\{u_i+J\}$, so that $K = K'\oplus J$ as vector spaces. To show that K' is a subalgebra of K, note that $y_i y_j = (u_i - w_{i1})(u_j - w_{j1}) = (u_i' + \Sigma_{k=2} e_k w_{ik})(u_j' + \Sigma_{k=2} e_k w_{jk}) = u_i' u_j' + \Sigma_{k=2} e_k(u_i' w_{jk} + w_{ik} u_j')$ (since $J^2=0$) $= u_i' u_j' + \Sigma_{k=2} e_k((u_i - w_i)w_{jk} + w_{ik}(u_j - w_j)) = \Sigma q_{ijk} u_k' + \Sigma_{k=2} e_k(u_i w_{jk} + w_{ik} u_j)$ (since $J^2 = 0$) $= \Sigma q_{ijk} y_k + \Sigma_{k=2} e_k z_k$ where $q_{ijk} y_k$ and z_k are in K. But $\{e_1 = 1, e_2, \ldots, e_r\}$ is also a K-basis of $E\otimes_Q K$ so that each $z_k = 0$ whence $y_i y_j = \Sigma q_{ijk} y_k$.

Finally assume that $k \geq 3$ is the smallest integer with $J^{(k)} = 0$. Then $\mathcal{J}(K/J^{(2)}) = J/J^{(2)}$ and $\mathcal{J}(K/J^{(2)})^{(2)} = 0$. By induction on k, $K/J^{(2)} = L/J^{(2)} \oplus J/J^{(2)}$ for some subalgebra $L/J^{(2)}$ of $K/J^{(2)}$ isomorphic to $(K/J^{(2)})/\mathcal{J}(K/J^{(2)}) \simeq K/J$. Thus L is a subalgebra of K with $K = L + J$. Since $L/J^{(2)}$ is semi-simple, $\mathcal{J}(L) = J^{(2)}$. Again by induction on k, $L = K' \oplus J^{(2)}$ for some subalgebra K' of L. Thus $K = K' + J$ with $K' \cap J = 0$, since $K' \simeq L/J^{(2)} \simeq L/\mathcal{J}(L)$ is semi-simple, so that $K = K' \oplus J$. ///

<u>Theorem 14.2 (Beaumont-Pierce [1])</u>: There is a subring T of R such that $T \oplus N(R)$ is a subgroup of R and $R/(T\oplus N(R))$ is finite.

<u>Proof</u>. Write $K = K' \oplus \mathcal{J}(K)$ as in Theorem 14.1 and let $T = R \cap K'$, a subring of R. Then $T \oplus N(R) \subseteq R$, since $T \oplus N(R) \subseteq K' \oplus \mathcal{J}(K)$, and $T \oplus N(R)$ is a subgroup of R. It suffices to prove that $R/T\oplus N(R)$ is finite.

Define $T_1 = \{x \in K' | x + y \in R$ for some $y \in J(K)\}$. Then $T \subseteq T_1 \subseteq K$, T_1 is a subring of K', and $\phi : R/(T \oplus N(R)) \to T_1/T$, defined by $\phi(r + T \oplus N(R)) = k + T$, where $r = k + j \in K' \oplus J(K)$, is a well-defined isomorphism. Thus, it is sufficient to prove that T_1/T is finite.

Now $K' = QT = QT_1$ is a semi-simple Q-algebra, since $K' \simeq K/J(K)$ as a Q-algebra, and $C(T) \subseteq C(T_1) \subseteq F' = C(K')$. Thus, T is a finitely generated $C(T)$-module and T_1 is a finitely generated $C(T_1)$-module (Theorem 9.10). It is now sufficient to prove that $C(T_1)/C(T)$ is finite, in which case $C(T_1)$, hence T_1, is a finitely generated $C(T)$-module so that T_1/T is finite.

Let p be a prime, $0 < \ell \in Z$ and define $I_\ell = (p^\ell C(T_1)) \cap C(T)$. Choose $0 < r \in Z$ with $J(K)^{(r)} = 0$. Let $p^\ell x_1 = z_1$, $p^\ell x_2 = z_2$ be elements of I_ℓ with $x_i \in C(T_1)$, $z_i \in C(T)$. Then $x_1 - y_1 = r_1$, $x_2 - y_2 = r_2$ for some $y_i \in J(K)$, $r_i \in R$. Now $z_1 z_2 - p^{2\ell} y_1 y_2 = z_1(z_2 - p^\ell y_2) + (z_1 - p^\ell y_1)z_2 - (z_1 - p^\ell y_1)(z_2 - p^\ell y_2) = z_1 p^\ell r_2 + p^\ell r_1 z_2 - p^{2\ell} r_1 r_2 \in p^\ell R$, since $z_i \in C(T) \subseteq R$, $r_i \in R$. By induction, if $p^\ell x_i = z_i \in I_\ell$ with $x_i \in C(T_1)$, $z_i \in C(T)$, $x_i - y_i = r_i \in R$ and $y_i \in J(K)$ then $z_1 z_2 \dots z_r - p^{r\ell} y_1 y_2 \dots y_r \in p^\ell R$. But $y_1 y_2 \dots y_r \in J(K)^{(r)} = 0$ so $z_1 z_2 \dots z_r \in p^\ell R \cap C(T) = p^\ell C(T)$ (since $T = R \cap K'$ is pure in R) and $(I_\ell)^{(r)} \subseteq p^\ell C(T)$.

There is $0 \neq n \in Z$ with $nS \subseteq C(T) \subseteq S = S_1 \times \dots \times S_k$, where each S_i is a Dedekind domain. Hence, $pS = P_1^{e_1} \dots P_m^{e_m}$, each $e_j \neq 0$ and each P_j is a maximal ideal of S. Moreover, $p^\ell S = P_1^{\ell e1} \dots P_m^{\ell em}$, $SI_\ell = P_1^{f_{\ell 1}} \dots P_m^{f_{\ell m}}$ and $(SI_\ell)^{(r)} \subseteq p^\ell SC(T) = p^\ell S$, so that $rf_{\ell j} \geq \ell e_j$. As $\ell \to \infty$, $f_{\ell j} \to \infty$, since r is fixed, hence there is some $0 < t \in Z$ such that if $\ell \geq t$, $f_{\ell j} \geq e_j$ for each $1 \leq j \leq m$. Thus, if $\ell \geq t$, $SI_\ell \subseteq pS$. But $S/C(T)$ is finite so that t can be chosen such that if $\ell \geq t$, then $I_\ell = p^\ell C(T_1) \cap C(T) \subseteq pC(T)$. It now follows that $(C(T_1)/C(T))_p$ is bounded by p^{t-1}, hence finite.

It now suffices to prove that if S is a Dedekind domain then S/pS is semi-simple for almost all primes p of Z; in which case $(S_1 \times \dots \times S_k)/p(S_1 \times \dots \times S_k)$, hence $C(T)/pC(T)$, is semi-simple for almost all primes p of Z (noting that if p does not divide n then $C(T)/pC(T) = S/pS = S_1/pS_1 \times \dots \times S_k/pS_k$). Now if $C(T)/pC(T)$ is semi-simple then $I_1 + pC(T)$ is an ideal of $C(T)/pC(T)$ so that $I_1 + pC(T) = I_1^{(r)} + pC(T)$, since ideals of semi-simple artinian algebras are idempotent, where $I_1 = (pC(T_1)) \cap C(T)$. As above, $I_1^{(r)} \subseteq pC(T)$ so that $I_1 \subseteq pC(T)$ and $(C(T_1)/C(T))_p = 0$. Consequently, $C(T_1)/C(T)$ must be finite as needed.

Finally, let S be a Dedekind domain and p be a prime of Z. Then S/pS is semi-simple iff p is unramified in S (i.e. $pS = P_1 P_2 \ldots P_\ell$ is a product of distinct prime ideals of S). But it is well known that almost all primes of Z are unramified in the Dedekind domain S (e.g. Zariski-Samuel [1]). ///

As a consequence of Theorem 14.2, R^+ is quasi-isomorphic to $T^+ \oplus N(R)^+$ where QT is a semi-simple Q-algebra. The next series of results describes the structure of T^+, up to quasi-isomorphism, in terms of additive groups of subrings of algebraic number fields, (due to Beaumont-Pierce [3]).

If QR is a simple algebra then a subfield F' of F is a field of definition of R if there is an F'-basis $\{r_1, \ldots, r_k\}$ of QR contained in R and $0 \neq n \in Z$ with $nR \subseteq (R \cap F')r_1 \oplus \ldots \oplus (R \cap F')r_k \subseteq R$. For example, $E_{QE_Z(R)}(QR)$ is a field of definition of R (Theorem 9.10). Fields of definition are not unique, for example if $R = Z[i]$ is the ring of Gaussian integers then Q and $QR = Q(i)$ are both fields of definition of $R = Z \oplus Zi$. In general, F is a field of definition of R, as a consequence of Theorem 9.10, noting that $C(R)$ is a domain, R is a finitely generated $C(R)$-module and $QC(R) = C(QR) = F$ with $C(R) = F \cap R$.

Theorem 14.3. Suppose that $K = QR$ is a simple Q-algebra. Then the following are equivalent:

(a) R^+ is strongly indecomposable;

(b) K is a field and is the smallest field of definition of R;

(c) The map $\phi : R \to E_Z(R)$, defined by $\phi(r)$ = left multiplication by r, is a ring isomorphism.

Proof. (a) => (b) If F' is a field of definition of R then $nR \subseteq (F' \cap R)r_1 \oplus \ldots \oplus (F' \cap R)r_k \subseteq R$ for some $0 \neq n \in Z$. Since R is strongly indecomposable, $k = 1$ and $Q(F' \cap R) = F' = QR = K$ is a field.

(b)⟹(c) As in the proof of Theorem 9.10, $QE_Z(R) \simeq E_D(QR)$ where $D \subseteq F$ is a field of definition of R. Thus, $D = F = QR$ and $QR \simeq E_{QR}(QR) \simeq QE_Z(R)$, the isomorphism from $QR \to QE_Z(R)$ being given by $x \to$ left multiplication by x. Consequently, ϕ is a 1-1 ring homomorphism. If $f \in E_Z(R)$ then f is left multiplication by $(m/n)r$ for some $m/n \in Q$, $r \in R$. Since $f(1) \in R$, $(m/n)r \in R$ so that $f = \phi((m/n)r)$ as desired.

(c)⟹(a) First of all, R is commutative: if $0 \neq x \in R$ then right multiplication by x is in $E_Z(R)$ so that there is $y \in R$ with $ys = sx$ for all $s \in R$. In particular, $y = y \cdot 1 = 1 \cdot x$ so that $x \in C(R)$ for each $x \in R$. Consequently, K is commutative. Since K is a simple Q-algebra, K is a field with $K \simeq QE_Z(R)$. Therefore, $E_Z(R)$ is a domain, hence R^+ is strongly indecomposable. ///

Corollary 14.4. If $N(R) = 0$ then there is $0 \neq n \in Z$ with $nR^+ \subseteq B_1 \oplus \dots \oplus B_k \subseteq R^+$, where each $B_i \simeq A_i^{n_i}$, A_i is the additive group of a subring of an algebraic number field, and A_i is strongly indecomposable.

Proof. In view of Theorem 9.9, it suffices to assume that $K = QR$ is a simple algebra. Let F' be a smallest field of definition of R so that R is quasi-isomorphic to $(F' \cap R)^m$ for some m. Then it follows that $F' = Q(F' \cap R)$ is the smallest field of definition of $F' \cap R$ so that $(F' \cap R)^+$ is strongly indecomposable (Theorem 14.3) and the additive group of a subring of F', an algebraic number field. ///

Example 14.5. Assume that $K = QR$ is a field of dimension m over Q, there is a prime p of Z with $pR = P_1 P_2^{e_2} \dots P_k^{e_k}$ a product of distinct prime ideals of R, $k > 1$ and that $\dim_{Z/pZ}(R/P_1) = \ell$ with g.c.d. $(\ell, m) = 1$. If R_{P_1} is the localization of R at the prime ideal P_1 then $R_{P_1}^+$ is strongly indecomposalbe and $E_Z(R_{P_1}) \simeq R_{P_1}$.

Proof. Note that $R_{P_1}/pR_{P_1} \simeq R_{P_1}/P_1 R_{P_1} \simeq R/P_1$. Since K is a field, R_{P_1} is quasi-isomorphic to A^n for some strongly indecomposable A (Corollary 14.4). Thus, $\text{p-rank}(R_{P_1}) = \ell = n(\text{p-rank}(A))$ and $\text{rank}(R_{P_1}^+) = m = n\,\text{rank}(A)$. Since g.c.d. $(m,\ell) = 1$, $n = 1$ so that $R^+_{P_1}$ is strongly indecomposable. Moreover, $E_Z(R_{P_1}) \simeq R_{P_1}$ by Theorem 14.3.

For example, if $R = Z[\sqrt{-5}]$ then $29R = \langle 3 + 2\sqrt{-5}\rangle\langle 3 - 2\sqrt{-5}\rangle = P_1P_2$, $2 = \dim_Q(QR)$ is relatively prime to $\dim_{Z/29Z}(R/P_1) = 1$, so that $R^+_{P_1}$ is strongly indecomposable. ///

The following theorem, due essentially to Beaumont-Pierce [3], gives a characterization of strongly indecomposable additive groups of subrings of algebraic number fields up to quasi-equality.

Theorem 14.6. Suppose that $S \subseteq QS = F$ is an algebraic number field and let $\overline{S}$ be the integral closure of S in F. Then S^+ is strongly indecomposable iff $\overline{S}$ is not the integral closure of S' in F for any subring $S' \subseteq QS' = F'$ such that F' is a proper subfield of F.

Proof. $(\Rightarrow)$ Assume that S^+ is strongly indecomposable. Then $\overline{S}/S$ is finite, by Corollary 10.12, so that $\overline{S}$ is strongly indecomposable. Suppose that $\overline{S}$ is the integral closure of some Dedekind domain S' with $QS' \subseteq F$. Then $\overline{S}$ is a finitely generated torsion free S'-module, by Theorem 10.11, so that $\overline{S}$ contains a free S'-module as an S'-submodule of finite index. Since $\overline{S}$ is strongly indecomposable, $\overline{S}/S'$ must be finite so that $QS' = Q\overline{S} = QS = F$.

$(\Leftarrow)$ As a consequence of Corollary 14.4 there is a Dedekind domain $S' \subseteq F$ such that S' is strongly indecomposable $T = S'x_1 \oplus \ldots \oplus S'x_k \subseteq S$ and S/T is finite where $F' = QS'$ is a smallest field of definition of S. Now S is an S'-order in F hence $\overline{S}$ is an integrally closed S'-order in F containing S'. Thus, $\overline{S}$ is the integral closure of S' in F, by Proposition 10.1, so that $QS' = F = QS$ by hypothesis. Thus, $\overline{S}/S'$ is finite so that $\overline{S}$, hence S, is strongly indecomposable. ///

The ring R is an E-ring if $\phi : R \to E_Z(R)$, given by $\phi(r)$ = left multiplication by r, is an isomorphism (Schultz [1]).

Corollary 14.7. The ring R is an E-ring iff there is $0 \neq n \in Z$ with $n(R_1 \times \ldots \times R_k) \subseteq R \subseteq R_1 \times \ldots \times R_k$, each R_i is a Dedekind domain, each R_i^+ is strongly indecomposable, and $\mathrm{Hom}_Z(R_i, R_j) = 0$ if $i \neq j$.

Proof. Suppose that $S \subseteq R$ is a ring inclusion with R/S finite. Then R is an E-ring iff $QR \to QE_Z(R)$ is an isomorphism (as in the proof of Theorem 14.3). Thus R is an E-ring iff S is an E-ring.

($\Leftarrow$) Each R_i is an E-ring by Theorem 14.3, hence $R_1 \times \ldots \times R_k$ is an E-ring since $E_Z(R_1 \times \ldots \times R_k) = E_Z(R_1) \times \ldots \times E_Z(R_k)$. Therefore, R is an E-ring by the preceding remarks.

($\Rightarrow$) If R is an E-ring then R is commutative, as in the proof of Theorem 14.3. By Theorem 14.2, $R/(T \oplus N(R))$ is finite for some subring T of R, say $nR \subseteq T \oplus N(R)$ for $0 \neq n \in Z$. Let $f : R \to N(R)$ be multiplication by n followed by a projection onto $N(R)$. Then f is left multiplication by some r, necessarily in $N(R)$. Since $n = n \cdot 1 \in T$, $f(n) = rn = 0$, noting that $T \subseteq \text{Kernel}(f)$. Hence $r = 0$, $f = 0$, and $N(R) = 0$.

Now $R \subseteq \overline{R} = R_1 \times \ldots \times R_k$ where $\overline{R}/R$ is finite and each R_i is a maximal order. Then $\overline{R}$ is an E-ring and $\overline{R} \simeq E_Z(\overline{R})$ is commutative. Therefore, each R_i is a Dedekind domain, R_i^+ is strongly indecomposable (Corollary 14.4), and $\text{Hom}_Z(R_i, R_j) = 0$ if $i \neq j$. ///

EXERCISES

14.1 Suppose that K is a finite dimension Q-algebra and that $K/J(K)$ is a product of matrix rings over a field E. Prove that K contains a subalgebra $L \simeq K/J(K)$. (Hint: Use Exercise 10.2 and Theorem 9.5).

§15. Q-simple and p-simple Groups:

A finite rank torsion free group A is Q-simple if $QE(A)$ is a simple algebra.

Theorem 15.1. Let A be a finite rank torsion free group. The following are equivalent:

(a) A is Q-simple;

(b) A is quasi-isomorphic to B^n for some $0 \neq n \in Z$, where B is a finite rank torsion free group and $QE(B)$ is a division algebra.

(c) If B is a fully invariant subgroup of A with $\mathrm{Hom}(A,B) \neq 0$ then A/B is finite.

Proof. (a)⟹(b) Write $QE(A) = I_1 \oplus \ldots \oplus I_n$ as a direct sum of irreducible right ideals with $I_i \simeq I_j$ for each i, j. Then $mA \subseteq B_1 \oplus \ldots \oplus B_n \subseteq A$ for some $0 \neq m \in Z$, where $QHom(A,B_i) = I_i$; B_i is quasi-isomorphic to B_j for each i, j; and each $QE(B_i)$ is a division algebra (Corollary 9.7).

(b)⟹(a) Since $QE(A) \simeq QE(B^n) \simeq Mat_n(QE(B))$ and $QE(B)$ is a division algebra, $QE(A)$ is simple.

(a)⟹(c) Let $I = \mathrm{Hom}(A,B)$, a non-zero ideal of $E(A)$. Then $QI = QE(A)$, since $QE(A)$ is simple. If $0 \neq m \in Z$ with $m \in I$ then $mA \subseteq IA \subseteq B \subseteq A$ and A/B is finite.

(c)⟹(a) Let I' be a non-zero ideal of $QE(A)$ so that $I = I' \cap E(A)$ is a non-zero ideal of $E(A)$. Now $B = IA$ is a fully invariant subgroup of A with $\mathrm{Hom}(A,B) \neq 0$ so that A/B is finite. In particular, $\mathcal{J}(QE(A)) = 0$ otherwise $A/N(E(A))A$ is finite and $QA = \mathcal{J}(QE(A))QA$ contradicting Nakayama's Lemma. Thus $QE(A)$ is a semi-simple algebra so $I' = QI = fQE(A)$ for some $f \in I$. Thus $QA = QB = fQA$ so that f is an automorphism of QA. Hence $fQE(A) = QI = I' = QE(A)$. ///

A finite rank torsion free group A is irreducible if whenever B is a pure non-zero fully invariant subgroup of A then $B = A$ (Reid [4]).

Theorem 15.2 (Reid [4]). The following are equivalent for a finite rank torsion free group A.

(a) A is irreducible;

(b) QA is an irreducible left $QE(A)$-module;

(c) $QE(A) \simeq Mat_m(D)$, where D is a division algebra with $rank(A) = m \cdot dim_Q(D)$;

(d) A is quasi-isomorphic to B^m for some $0 < m \in Z$, where B is a strongly indecomposable irreducible group.

Proof. There is a 1-1 correspondence from pure fully invariant subgroups of A to $QE(A)$-submodules of QA given by $B \to QB$ with inverse $M \to M \cap A$, whence (a)<=>(b).

(b) =>(c) Since QA is an irreducible $QE(A)$-module $J(QE(A)) = 0$, otherwise $J(QE(A)QA = QA$ which is impossible by Nakayama's Lemma. Moreover, $QE(A)$ is a simple algebra, otherwise $QE(A) = K_1 \times K_2$ is the product of non-zero algebras so that $QA = K_1QA \oplus K_2QA$ where each $K_iQA \neq 0$, a contradiction. Thus, $QE(A) \simeq (QA)^m$, hence $QE(A) \simeq Mat_m(E_{QE(A)}(QA))$ with $E_{QE(A)}(QA) = D$ a division algebra. Finally, $dim_Q(QE(A)) = m\ dim_Q(QA) = m^2 dim_Q D$ so that $rank(A) = dim_Q(QA) = m\ dim_Q D$.

(c) =>(d) As a consequence of Theorem 15.1, A is quasi-isomorphic to B^m where B is a strongly indecomposable group with $QE(B) \simeq D$. Furthermore, $rank(B) = dim_Q(D)$ since $m(rank(B)) = rank(A) = m(dim_Q(D))$. To see that B is irreducible, let C be a non-zero pure fully invariant subgroup of B and $0 \neq x \in C$. Then $E(B)x \subseteq C$ with $E(B)x \simeq E(B)$ since $QE(B)$ is a division algebra. But $rank(E(B)) = dim_Q(QE(B) = dim_Q(D) = rank(B)$ so that $C = B$.

(d) =>(a) Choose $0 \neq k \in Z$ with $kA \subseteq B_1 \oplus \ldots \oplus B_m \subseteq A$ where each $B_i \simeq B$ is strongly indecomposable and irreducible. Let C be a non-zero pure fully invariant subgroup of A and $0 \neq x \in C$ so that $E(A)x \subseteq C$. Now $kx = b_1 + \ldots + b_m$ for some $b_i \in B$ and $E(B_1 \oplus \ldots \oplus B_m)(b_1 + \ldots + b_m) \subseteq kE(A)x$. Since $0 \neq x$, some $b_i \neq 0$. But B is irreducible so $rank(A) = rank(B_1 \oplus \ldots \oplus B_m) \leq rank(E(B_1 \oplus \ldots \oplus B_m)(b_1 + \ldots + b_m))$. Thus $rank(kE(A)x) = rank(E(A)x) = rank(A)$ so that $A = C$. ///

If A is strongly indecomposable then A is Q-simple iff $QE(A)$ is a division algebra (Theorem 15.1) while A is irreducible iff $QE(A)$ is a division algebra with $\dim_Q(QE(A)) = \text{rank}(A)$ (Theorem 15.2).

Corollary 15.3. Suppose that A is a torsion free group of finite rank.

(a) If A is irreducible then A is homogeneous.

(b) A is an irreducible almost completely decomposable group iff A is homogeneous completely decomposable.

Proof. (a) If $\tau \in \text{typeset}(A)$ then $A(\tau) = \{x \in A \mid \text{type}_A(x) \geq \tau\}$ is a non-zero pure fully invariant subgroup, hence $A(\tau) = A$.

(b) ($\Leftarrow$) Apply Theorem 15.2.

($\Rightarrow$) Apply (a) and Theorem 2.3. ///

The preceding properties of Q-simple and irreducible groups were essentially derived from the fact that $QE(A)$ is an artinian algebra. If p is a prime of Z then $E(A)/pE(A)$ is an artinian algebra. Hence define a finite rank torsion free group A to be p-simple if $E(A)/pE(A)$ is a simple algebra and p-irreducible if A/pA is an irreducible left $E(A)/pE(A)$-module.

Proposition 15.4. Let A be a finite rank torsion free group. Then A is p-simple for each prime p of Z iff whenever I is an ideal of $E(A)$ with $E(A)/I$ finite then $I = nE(A)$ for some $n \in Z$.

Proof. ($\Leftarrow$) If $I/pE(A)$ is an ideal of $E(A)/pE(A)$ then $I/pE(A) = 0$ or $E(A)/pE(A)$ so that $E(A)/pE(A)$ is simple.

($\Rightarrow$) Let $0 < n$ be the least integer with $nE(A) \subseteq I$ and let p be a prime divisor of n. Then $I + pE(A) = E(A)$ or $I + pE(A) = pE(A)$ since $E(A)/pE(A)$ is simple. In the first case $(n/p)E(A) \subseteq I$, contradicting the minimality of n. In the second case, $(n/p)E(A) \subseteq (1/p)I \subseteq E(A)$ so $(1/p)I = (n/p)E(A)$ by induction on n. ///

Theorem 15.5. Let A be a reduced torsion free group of finite rank. Then A is p-simple for each prime p of Z iff $A = A_1 \oplus \ldots \oplus A_k$ where each A_i is fully invariant in A, each A_i is Q-simple and p-simple for each prime p, and if p is a prime then there is some j with $A/pA = A_j/pA_j$.

Proof. (⇐) Note that $E(A) = E(A_1) \times \ldots \times E(A_k)$ and if p is a prime of Z then $E(A)/pE(A) = E(A_j)/pE(A_j)$ is simple where $A/pA = A_j/pA_j$, since $pA_i = A_i$ hence $pE(A_i) = E(A_i)$, for $i \neq j$.

(⇒) For each prime p, $(N(E(A)) + pE(A))/pE(A) \subseteq J(E(A)/pE(A)) = 0$. Thus, $N(E(A)) \subseteq pE(A) \cap N(E(A)) = pN(E(A))$ for each p. Since A is reduced, $E(A)$ is reduced, hence $N(E(A)) = 0$.

Let $R = E(A)$ so that $\overline{R}/R$ is finite where $\overline{R} = R_1 \times \ldots \times R_k$, each R_i is a maximal S_i-order in the simple algebra QR_i and S_i is the integral closure of $C(R_i)$ in the algebraic number field $QC(R_i)$ (Corollary 10.14). Now $I = n\overline{R}$ is an ideal of R for some $0 \neq n \in Z$ so that $I = mR$ for some $m \in Z$ (Proposition 15.4). Hence, $mR = I = \overline{R}I = m\overline{R}$ so that $R = \overline{R}$.

Define $A_i = R_iA$ so that $A = A_1 \oplus \ldots \oplus A_k$, $E(A_i) = R_i$, and $E(A) = R_1 \times \ldots \times R_k$. Each A_i is Q-simple since QR_i is a simple algebra. If p is a prime then $E(A)/pE(A) = E(A_1)/pE(A_1) \times \ldots \times E(A_k)/pe(A_k)$ is simple so there is some j with $E(A)/pE(A) = E(A_j)/pE(A_j)$, whence $pE(A_i) = E(A_i)$ and $pA_i = A_i$ if $i \neq j$. Thus, $A/pA = A_j/pA_j$. ///

Corollary 15.6. Suppose that A is Q-simple and p-simple for each prime p of Z and let $S = C(E(A))$.

(a) S is a principal ideal domain such that every element of S is a rational integral multiple of a unit of S.

(b) $E(A)$ is a maximal S-order in $QE(A)$.

(c) For some $0 < k \in Z$, $E(A) \simeq S^k$ as S-modules.

Proof. (a) Note that S is a domain hence if $0 \neq s \in S$ then $sE(A)$ is an ideal of $E(A)$ with $E(A)/sE(A)$ finite. Thus $sE(A) = nE(A)$ for some $n \in Z$ (Proposition 15.4). Consequently, $s = nu$ for some unit u of S.

(b) is a consequence of Proposition 15.4 and (c) is a consequence of (a) and (b). ///

Corollary 15.7. Let A and B be finite rank torsion free groups.

(a) If A and B are quasi-isomorphic and A is Q-simple then B is Q-simple.

(b) If A and B are nearly isomorphic and if A is p-simple then B is p-simple.

Proof. If $0 \neq n \in Z$ with $nA \subseteq B \subseteq A$ then $QE(A) \simeq QE(B)$ while if p is a prime not dividing n then $E(A)/pE(A) \simeq E(B)/pE(B)$. ///

Theorem 15.8. Suppose that A is a torsion free group of finite rank. Then A is Q-simple and p-simple for each prime p of Z iff $A \simeq B^k \oplus B'$ for some $0 < k \in Z$ where B is strongly indecomposable, Q-simple, p-simple for each prime p of Z, and B' is nearly isomorphic to B. Consequently, A is indecomposable iff A is strongly indecomposable.

Proof. ($\Leftarrow$) By Corollary 15.7, it is sufficient to assume that $A = B^{k+1}$. Thus, $QE(A) \simeq Mat_{k+1}(QE(B))$ is simple since $QE(B)$ is a division algebra. Furthermore, $E(B)/pE(B) \simeq Mat_{m_p}(F_p)$ for some finite field F_p whence $E(A)/pE(A) \simeq Mat_{(k+1)m_p}(F_p)$ is a simple algebra for each prime p of Z.

($\Rightarrow$) By Theorem 15.1, there is $0 \neq m \in Z$ with $mA \subseteq B_1 \oplus \dots \oplus B_{k+1} \subseteq A$ where each B_i is quasi-isomorphic to a strongly indecomposable Q-simple group B'.

Let X be the pure subgroup of A generated by $\{B_i \mid i \neq 1\}$ and let $B = A/X$. Then B is quasi-isomorphic to B_1 and $Hom(A,B)A = B$. It now suffices to assume that $mA \subseteq C \subseteq A$ for some $0 \neq m \in Z$ where $C = B_1 \oplus \dots \oplus B_{k+1}$, each $B_i \simeq B$ is strongly indecomposable and $Hom(A,C)A = C$ (replace each B_i by a subgroup of finite index isomorphic to B).

Now $E(A)$ is a maximal S-order, $S = C(E(A))$ is a principal ideal domain by Theorem 15.6, and $Hom(A,C)$ is an $E(A)$-lattice with $QHom(A,C) \simeq QE(A)$ as $QE(A)$-modules. Thus $Hom(A,C)$ and $E(A)$ are in the same genus class (Corollary 12.4) and $Hom(A,C)$ is a finitely generated projective R-module (Proposition 12.1). Thus, A and C are nearly isomorphic by Corollary 12.7, noting that $Hom(A,C)A = C$. Consequently, A is nearly isomorphic to B^{k+1} so that $A \simeq B^k \oplus B'$ for some B' nearly isomorphic to B (Corollary 12.9). Moreover, B^{k+1} is p-simple for each prime p of Z by Corollary 15.7 from which it follows that B is p-simple for each prime p of Z. ///

Theorem 15.9. The following are equivalent for a finite rank torsion free group A.

(a) A is p-irreducible;

(b) If $pA \subseteq B$ is a fully invariant subgroup of A then $B = A$ or $B = pA$;

(c) $E(A)/pE(A) \simeq \mathrm{Mat}_{m_p}(F_p)$ where F_p is a finite field and p-rank(A) = m_pp-rank(F_p).

Proof. Imitate the proof of Theorem 15.2.

Corollary 15.10. Suppose that A and B are finite rank torsion free groups.

(a) If A is quasi-isomorphic to B and if A is irreducible then B is irreducible.

(b) If A is nearly isomorphic to B and if A is p-irreducible then B is p-irreducible.

Proof. As in Corollary 15.7, using Theorem 15.2 and Theorem 15.9.

Corollary 15.11. Let A be a reduced finite rank torsion free group.

(a) A is p-irreducible for each prime p of Z iff $A = A_1 \oplus \cdots \oplus A_k$ where $\mathrm{Hom}(A_i, A_j) = 0$ if $i \neq j$, each A_i is Q-simple and p-irreducible for each prime p of Z, and if p is a prime then there is some j with $A/pA = A_j/pA_j$.

(b) A is Q-simple (irreducible) and p-irreducible for each prime p of Z iff $A \simeq B^k \oplus B'$ where B is strongly indecomposable, Q-simple (irreducible), p-irreducible for each prime p of Z, and B' is nearly isomorphic to B.

Proof. Apply Theorem 15.5, Theorem 15.8, Theorem 15.2, and Theorem 15.9.

A finite rank torsion free group is a Murley group if p-rank(A) ≤ 1 for each prime p of Z.

Corollary 15.12 (Murley [1]). Suppose that A is a finite rank torsion free Murley group.

(a) If B is a torsion free group quasi-isomorphic to A then B is isomorphic to A.

(b) $A = A_1 \oplus \cdots \oplus A_k$ where each A_i is an indecomposable Murley group.

(c) A is p-irreducible for each prime p of Z.

(d) If A is indecomposable then $E(A)$ is a principal ideal domain such that every element of $E(A)$ is an integral multiple of a unit of $E(A)$.

(e) If $A = A_1 \oplus \ldots \oplus A_m = B_1 \oplus \ldots \oplus B_n$ with each A_i and B_j indecomposable then $m = n$ and $A_i \simeq B_{\sigma(i)}$ for some permutation σ of m^+.

Proof. (a) Follows from the fact that if $0 \neq n \in Z$ then A/nA is cyclic so that if $nA \subseteq B$ then $B = mA$ for some $m \in Z$.

(b) Write $A = A_1 \oplus \ldots \oplus A_k$ as a direct sum of indecomposable groups and note that $p\text{-rank}(A_i) \leq p\text{-rank}(A) \leq 1$ for each prime p of Z.

(c) Assume that p is a prime with $pA \neq A$ and let $x + pA$ be a cyclic generator of A/pA. Define $\phi : E(A)/pE(A) \to A/pA$ by $\phi(f+pE(A)) = f(x) + pA$, an epimorphism. If $f + pE(A) \in \text{Kernel}(\phi)$ then $f(x) \in pA$, whence $f(A) \subseteq pA$ and $f \in pE(A)$. Thus, $E(A)/pE(A) \simeq Z/pZ$ so that Theorem 15.9 applies.

(d) If A is indecomposable then A is strongly indecomposable by Corollary 15.11 and (c). Hence $E(A)$ is a maximal $C(E(A))$-order in the simple algebra $QE(A)$, $C(E(A))$ is a principal ideal domain such that every element of $C(E(A))$ is an integral multiple of a unit of $C(E(A))$ and $E(A) \simeq C(E(A))^k$ for some k (Corollary 15.6). But $E(A)/pE(A) \simeq Z/pZ$, as in (c), so that $k = 1$ and $E(A) \simeq C(E(A))$.

(e) is a consequence of Corollary 7.9 and (a). Alternatively, apply Corollary 6.7, Corollary 5.2, (d) and (a). ///

Note that every rank-1 group is a Murley group.

Example 15.13. A reduced finite rank torsion free group is a Murley group iff A is isomorphic to a pure subgroup of $\pi_p Z_p^*$, where Z_p^* is the p-adic integers.

Proof. ($\Leftarrow$) For each p, $p\text{-rank}(A) \leq p\text{-rank}(\pi_p Z_p^*) = p\text{-rank}(Z_p^*) = 1$.

($\Rightarrow$) Since A is reduced, A is isomorphic to a pure subgroup of $\pi_p A_p^*$, where A_p^* is the p-adic completion of A, i.e., $A_p^* = \varprojlim A/p^i A$. Since $p\text{-rank}(A) \leq 1$, $A_p^* \simeq Z_p^*$, if $pA \neq A$, or 0 if $pA = A$.

A finite rank torsion free group A is strongly homogeneous if whenever X and Y are pure rank-1 subgroups of A then there is an automorphism u of A with $u(X) = Y$.

Theorem 15.14. Let A be a finite rank torsion free group. Then the following are equivalent:

(a) A is irreducible, p-irreducible for each prime p of Z and $QE(A) \simeq Mat_m(F)$ where F is a field;

(b) $A \simeq (S \otimes_Z X)^m$ where X is a torsion free group of rank 1 and S is a principal ideal domain such that every element of S is an integral multiple of a unit of S;

(c) A is strongly homogeneous.

Proof. (a) $\Rightarrow$ (b) Now $A \simeq B^{m-1} \oplus B'$, where B is strongly indecomposable, irreducible, p-irreducible for each prime p of Z and B' is nearly isomorphic to B (Corollary 15.11). Moreover, $QE(B) \simeq F$ with $\mathrm{rank}(E(B)) = \dim_Q(F) = \mathrm{rank}(B)$ by Theorem 15.2. Also $E(B) = \mathcal{C}(E(B))$ is a principal ideal domain such that every element is an integral multiple of a unit (Corollary 15.6) so that $B' \simeq B$ (Corollary 12.10). Consequently, $A \simeq B^m$.

Let X be a pure rank-1 subgroup of B and define $\phi : E(B) \otimes_Z X \to B$ by $\phi(f \otimes x) = f(x)$. It is sufficient to prove that ϕ is into, in which case ϕ is an isomorphism, since $\mathrm{rank}(B) = \mathrm{rank}(E(B)) = \mathrm{rank}(E(B) \otimes_Z X)$, and $A \simeq (E(B) \otimes_Z X)^m$.

Now $B/\mathrm{Image}(\phi)$ is torsion since $QE(B)$ is a field and $\mathrm{rank}(E(B)) = \mathrm{rank}(B)$. Let $b \in B$ and $pb = f(x)$ for a prime p, $x \in X$, and $f \in E(B)$, noting that since $\mathrm{rank}(X) = 1$ every element of $E(B) \otimes_Z X$ has form $f \otimes x$ for $f \in E(B)$, $x \in X$. But $f = nu$ for some $n \in Z$, u a unit of $E(B)$ so $pb = f(x) = nu(x)$. If p divides n then $b \in \mathrm{Image}(\phi)$. Otherwise $u(x)/p \in B$ so that $x/p = u^{-1}(u(x)/p) \in B$, hence X, and $b = f(x/p) \in \mathrm{Image}(\phi)$.

(b) $\Rightarrow$ (c) It suffices to assume that X is a pure rank 1 subgroup of A since if $S' = \langle 1 \rangle_*$ is a pure rank 1 subgroup of S then $S' \otimes_Z X$ is a pure rank 1 subgroup of $S \otimes_Z X$ and $S \otimes_Z (S' \otimes_Z X) \simeq (S \otimes_Z S') \otimes_Z X \simeq S \otimes_Z X$ (noting that S' is a subring of S).

It is sufficient to prove that if Y is a pure rank 1 subgroup of A then there is an automorphism u of A with $u(Y) = X$, since if $u'(Y') = X$ for another pure rank-1 subgroup Y' of A and u' an automorphism of A then $u^{-1}u'(Y') = Y$ and $u^{-1}u'$ is an automorphism of A.

Let $B = SY$. Then B is an S-pure submodule of A with S-rank$(B) = 1$ since every element of S is an integral multiple of a unit of S. Moreover, A is a finite rank homogeneous completely decomposable S-module. Consequently, B is an S-summand of A isomorphic to $S \otimes X$ (as in Exercise 1.4 using S in place of Z).

Now S is homogeneous, since every element of S is an integral multiple of a unit of S, so that $S \otimes X$ is homogeneous of type = type(X) (Exercise 2.1) whence $Y \simeq X$. If $f : X \to Y$ is an isomorphism then $1 \otimes f : S \otimes X \to S \otimes Y$ is an isomorphism. But $S \otimes Y \to B = SY$ is onto hence an isomorphism so there is an isomorphism $g : S \otimes X \to S \otimes Y$ with $g(X) = Y$. Since $S \otimes X$ and $S \otimes Y$ are summands of A, there is an automorphism u of A with $u(X) = Y$.

(c) $\Rightarrow$(a) Let $0 \neq B$ be a fully invariant subgroup. If B is pure in A, X a pure rank-1 subgroup of B and Y a pure rank-1 subgroup of A then there is an automorphism u of A with $u(X) = Y \subseteq B$. Thus $B = A$ so that A is irreducible.

Assume that $pA \subsetneq B$ for some prime p of Z, $0 \neq b \in B \backslash pA$ $X = \langle b \rangle_*$, $a \in A \backslash pA$, and $Y = \langle a \rangle_*$. Then $u(X) = Y$ for some automorphism u of A, i.e., $mu(b) = na$ for some relatively prime integers m and n. Hence $na \in B$. If $p|n$ then $b \in pA$ a contradiction. Otherwise write $1 = rn + sp$ so that $a = rna + spa \in B$. Thus $B = A$ and A is p-irreducible.

By Corollary 15.11, $A \simeq B^{m-1} \oplus B'$ where B is strongly indecomposable, irreducible, p-irreducible for each p and B' is nearly isomorphic to B. Since A is strongly homogeneous B is strongly homogeneous e.g., if X is a pure rank 1 subgroup of B and Y a pure rank 1 subgroup of B' then there is an automorphism u of A with $u(X) = Y$. Hence $\pi_B \mu \pi_B : B \to B$ is an isomorphism, noting that since B is strongly indecomposable and irreducible every non-zero endomorphism of B is a monomorphism so that $\mu : B \to B$.

Thus $A \simeq B^m$ and $QE(A) \simeq \text{Mat}_m(QE(B))$ so it suffices to prove that the division algebra $QE(B)$ is a field. If X is a pure rank 1 subgroup of B then $E(B) \otimes_Z X \to B$ is an isomorphism since B is strongly indecomposable, irreducible and strongly homogeneous. Hence $E(B)$ is strongly indecomposable. By Corollary 15.6(c) $E(B) = C(E(B))$ is commutative so that $QE(B)$ is a field. ///

If A is a finite rank homogeneous completely decomposable group then A is strongly homogeneous by the preceding theorem. The following corollary demonstrates that strongly homogeneous groups have many of the properties of homogeneous completely decomposable groups.

Corollary 15.15. Suppose that A and B are strongly homogeneous torsion free groups of finite rank.

(a) $A \simeq C^n$ for some $0 < n \in Z$, where C is strongly indecomposable, strongly homogeneous and $E(C)$ is a principal ideal domain such that every element is an integral multiple of a unit.

(b) If A is strongly indecomposable and G is a summand of A^n then $G \simeq A^m$ for some $0 < m \in Z$.

(c) A and B are isomorphic iff $\mathrm{rank}(A) = \mathrm{rank}(B)$, $\mathrm{type}(A) = \mathrm{type}(B)$, and $\mathcal{C}(E(A)) \simeq \mathcal{C}(E(B))$.

(d) A and B are quasi-isomorphic iff A and B are isomorphic.

Proof. (a) and (c) follow from Theorem 15.14 while (b) is a consequence of (a) and Corollary 5.2.

(d) follows from (c) and the fact that if A and B are quasi-isomorphic then $\mathrm{rank}(A) = \mathrm{rank}(B)$, $\mathrm{type}(A) = \mathrm{type}(B)$, and $\mathcal{C}(E(A))$, $\mathcal{C}(E(B))$ are quasi-isomorphic principal ideal domains, hence isomorphic as rings. ///

Corollary 15.16. Assume that A is a finite rank torsion free group. The following are equivalent.

(a) A is homogeneous completely decomposable;

(b) A is irreducible, p-irreducible for each prime p of Z, and $QE(A) \simeq \mathrm{Mat}_m(Q)$ for some $0 < m \in Z$;

(c) $\mathrm{rank}(E(A)) = \mathrm{rank}(A)^2$.

Proof. (a)$\Leftrightarrow$(c) is Corollary 1.13.

(a)$\Leftrightarrow$(b) is a consequence of Theorem 15.14. ///

EXERCISES

15.1 (Arnold [6]). Let S be a subring of QS, an algebraic number field. Then every element of S is a rational integral multiple of a unit of S iff $S = \cap\{J_P \mid P \in \mathcal{J}\}$ where J is the ring of algebraic integers of QS, $\mathcal{J}$ is a set of prime ideals of J, J_P is the localization of J at P, and if p is a prime of Z with $pJ = P_1^{e_1} P_2^{e_2} \ldots P_n^{e_n}$ a product of powers of distinct prime ideals of J then at most one $P_i \in \mathcal{J}$ and if $P_i \in \mathcal{J}$ then $e_i = 1$.

List of References

Armstrong, J.W.

[1] On p-pure subgroups of the p-adic integers, Topics in Abelian Groups, 315-321 (Chicago, Illinois, 1963).

[2] On the indecomposability of torsion-free abelian groups, Proc. Amer. Math. Soc. 16(1965), 323-325.

Arnold, D.M.

[1] A duality for torsion free modules of finite rank over a discrete valuation ring, Proc. London Math. Soc., (3) 24 (1972), 204-216.

[2] A duality for quotient divisible abelian groups of finite rank, Pacific J. Math., 42 (1972), 11-15.

[3] Exterior powers and torsion free modules over discrete valuation rings, Trans. Amer. Math. Soc., 170 (1972), 471-481.

[4] Algebraic K-theory of torsion free abelian groups, Symposia Mathematica (Academic Press, 1974), Proc. Conf. Abelian Groups, Rome, 1972, 179-193.

[5] A class of pure subgroups of completely decomposable abelian groups, Proc. Amer. Math. Soc., 41 (1973), 37-44.

[6] Strongly homogeneous torsion free abelian groups of finite rank, Proc. Amer. Math. Soc. 56 (1976), 67-72.

[7] Genera and decompositions of torsion free modules, Springer-Verlag Lecture Notes #616 (1977), 197-218.

[8] Endomorphism rings and subgroups of finite rank torsion free abelian groups, preprint

Arnold, D.M., and Lady, L.

[1] Endomorphism rings and direct sums of torsion free abelian groups, Trans. Amer. Math. Soc., 211 (1975), 225-237.

Arnold, D.M., and Murley, C.E.

[1] Abelian groups A, such that Hom(A, -) preserves direct sums of copies of A, Pacific J. Math., 56 (1975), 7-20.

Arnold, D.M.- Vinsonhaler, C., and Wickless, W.

[1] Quasi-pure projective and injective torsion free abelian groups of rank 2, Rocky Mountain J. Math., 6 (1976), 61-70.

Arnold, D.M., O'Brien, B., and Reid, J.,

[1] Quasi-pure projective and injective torsion free abelian groups Proc. London Math. Soc. (?) 38 (1978).

Arnold, D.M., Hunter, R., and Walker E.

[1] Summands of direct sums of cyclic valuated groups Symposia Mathematica, [Academic Press, 1979] , Proc. Conf. Abelian Groups, Rome, 1977.

Arnold, D.M., Hunter, R., and Richman, F.

[1] Global Azumaya theorems in additive categories, J. Pure and Appl. Algebra 16 (1980), 223-242.

Arnold, D.M., Pierce, R.S., Reid, J.D., Vinsonhaler, C., and Wickless, W.J.

[1] Torsion free abelian groups of finite rank projective as modules over their endomorphism rings, preprint.

Arsinov, M.N

[1] On the projective dimension of torsion-free abelian groups over the ring of its endomorphism [Russian] , Mat. Zametki 7 (1970), 117-124.

Baer, R.

[1] Types of elements and characteristic subgroups of abelian groups Proc. London Math. Soc. 39 (1935), 481-514.

[2] Abelian groups without elements of finite order, Duke Math. J. 3 (1937), 68-122.

Bass, H.

[1] Algebraic K-theory, Benjamin, New York, 1968.

[2] Torsion free and projective modules, Trans. Amer. Math. Soc. (1962), 319-327.

Beaumont, R.A., and Pierce, R.S.

[1] Torsion-free rings, Illinois. J. Math.5 (1961), 61-98.

[2] Torsion free groups of rank two, Mem. Amer. Math. Soc. #38 (1961).

[3] Subrings of algebraic number fields, Acta Sci. Math. (Szeged) 22 (1961), 202-216.

Beaumont, R.A., and Lawver, D.A.

[1] Srongly semi-simple abelian groups, Pacific J. Math. 48 ( ),

Beaumont, R.A., and Wisner, R.J.

[1] Rings with additive group which is torsion-free group of rank two, Acta Sci. Math. (Szeged) 20 (1959), 105-116.

Beaumont, R.A., and Zuckerman, H S.

[1] A characterization of the subgroups of the additive rationals, Pacific J. Math. 1 (1951), 169-177.

Bowshell, R.A., and Schultz, P.

[1] Unital rings whose additive endomorphisms commute, Math. Ann. 228 (1977), 197-214.

Brenner, S., and Butler, M.C.R.

[1] Endomorphism rings of vector spaces and torsion free abelian groups, J. London Math. Soc. 40 (1965), 183-187.

Butler, M.C.R.

[1] A class of torsion-free abelian groups of finite rank, Proc. London Math. Soc. 15 (1965), 680-698.

[2] On locally torsion-free rings of finite rank, J. London Math. Soc. 43 (1968), 297-300.

Campbell, M. O´N.

[1] Countable torsion-free abelian groups, Proc. London Math. Soc. 10 (1960), 1-23.

Charles, B.

[1] Sous-groupes fonctoriels et topologies, Studies on Abelian Groups, 75-92 (Paris, 1968).

Chase, S.U.

[1] Locally free modules and a problem of Whitehead, Illinois. J. Math. 6 (1962), 682-699.

Claborn, L.

[1] Every abelian group is a class group, Pacific J. Math. 18 (1966), 219-222.

Cornelius E.F., Jr.

[1] Note on quasi-decompositions of irreducible groups, Proc. Amer. Math. Soc. 26 (1970), 33-36.

Corner, A.L.S.

[1] A note on rank and direct decompositions of torsion-free abelian groups, Proc. Cambridge Philos. Soc. 57 (1961), 230-233, and 66 (1969), 239-240.

[2] Every countable reduced torsion-free ring is an endomorphism ring, Proc. London Math. Soc. 13 (1963), 687-710.

[3] Endomorphism rings of torsion-free abelian groups, Proc. Conf. Theory Groups, 59-69 (New York, London, Paris, 1967).

Corner, A.L.S., and Crawley, P.

[1] An abelian p-group without the isomorphic refinement property, Bull. Amer. Math. Soc. 74 (1968), 743-746.

Crawley, P., and Jonsson, B.

[1] Refinements for infinite direct decompositions of algebraic systems, Pacific J. Math. 14 (1964), 797-855.

Cruddis, T.B.

[1] On a class of torsion free abelian groups, Proc. London Math. Soc. 21 (1970), 243-276.

Douglas, A.J., and Farahat, H.K.

[1] The homological dimension of an abelian group as a module over its ring of endomorphisms, Monatsh. Math. 69 (1965), 294-305.

Dubois, D.W.

[1] Cohesive groups and p-adic integers, Publ. Math. Debrecen 12 (1965), 51-58.

Estes, D., and Ohm, J.

[1] Stable range in commutative rings, J. Alg. 7 (1967), 343-362.

Friedlander, E.

[1] Extension functions for rank-2 torsion free abelian groups, Pacific J. of Math, 58 (1975), 371-380.

Fuchs, L.

[1] Abelian Groups. Publ. House of the Hungar. Acad. Sci., Budapest, 1958.

[2] Recent results and problems on abelian groups, Topics in Abelian Groups, 9-40 (Chicago, Illinois, 1963).

[3] Note on direct decompositions of torsion-free abelian groups, Comment. Math. Helv. 46 (1971), 87-91.

[4] Some aspects of the theory of torsion-free abelian groups, Texas Tech. University Series. No. 9 , 32-58 (Lubbock, 1971)

[5] The cancellation property for modules, Lectures on Rings and Modules, Springer-Verlag Lecture Notes #246, 1970.

[6] Infinite Abelian Groups, Vol. I, Academic Press, 1970.

[7] Infinite Abelian Groups, Vol. II, Academic Press, 1973.

Fuchs, L. and Grabe, P.

[1] Number of indecomposable summands in direct decompositions of torsion free abelian groups, Rend. Sem. Mat. Univ. Padova 53 (1975), 135-148.

[2] On the cancellation of modules in direct sums over Dedekind domains, Indagationes Math. 33 (1971), 163-169.

Fuchs, L., and Loonstra, F.

[1] On direct decompositions of torsion-free abelian groups of finite rank, Rend. Sem. Mat. Univ. Padova 44 (1970), 75-83.

[2] On the cancellation of modules in direct sums over Dedekind domains, Indagationes Math. 33 (1971), 163-169.

Gardner, B.J.

[1] A note on types, Bull. Austral. Math. Soc. 2 (1970), 275-276.

Goodearl, K.R.

[1] Power cancellation of groups and modules, Pacific J. Math., 64 (1976), 387-411.

Griffith, P.

[1] Purely indecomposable torsion-free groups, Proc. Amer. Math Soc. 18 (1967), 738-742.

[2] Decomposition of pure subgroups of torsion free groups, Illinois J. Math. 12 (1968) 433-438.

[3] Infinite Abelian Groups, Chicago Lectures in Mathematics (Chicago and London, 1970).

de Groot, J.

[1] An isomorphism criterion for completely decomposable abelian groups, Math. Ann. 132 (1956), 328-332.

[2] Indecomposable abelian groups, Proc. Ned. Akad. Wetensch. 60 (1957), 137-145.

[3] Equivalent abelian groups, Canad. J. Math. 9 (1957), 291-297.

Hallett, J.T., and Hirsch, K.A.

[1] Torsion-free groups having finite automorphism groups, J. Algebra 2 (1965) 287-298.

[2] Die Konstruktion von Gruppen mit vorgeschriebenen Automorphismengruppen, J. Reine Angew. Math. 241 (1970), 32-46.

Hirsch, K.A., and Zassenhaus, H.

[1] Finite automorphism groups of torsion-free groups, J. London Math. Soc. 41 (1966), 545-549.

Hsiang, W.C., and Hsiang, W.

[1] Those abelian groups characterized by their completely decomposable subgroups of finite rank, Pacific J. Math. 11 (1961), 547-558.

Hungerford, T.

[1] Algebra, Holt, Rinehart and Winston, 1974.

Hunter, R. and Richman, F.

[1] Global Warfield groups, preprint.

Jacobinski, H.

[1] Genera and decompositions of lattices over orders, Acta. Math. 121 (1968), 1-29.

Jesmanowicz, L.

[1] On direct decompositions of torsion-free abelian groups, Bull. Acad. Polon. Sci. 8 (1960), 505-510.

Jonsson, B.

[1] On direct decompositions of torsion-free abelian groups, Math. Scand. 5 (1957), 230-235.

[2] On direct decomposition of torsion-free abelian groups, Math. Scand. 7 (1959), 361-371.

Kaplansky, I.

[1] Infinite Abelian Groups, University of Michigan press, Ann Arbor, Michigan, 1954, 1969.

[2] Commutative Rings, Allyn and Bacon, 1970.

Koehler, J.

[1] Some torsion-free rank two groups, Acta. Sci. Math. Szeged 25 (1964), 186-190.

[2] The type set of a torsion-free group of finite rank, Illinois J. Math. 9 (1965), 66-86.

Kolettis, G., Jr.

[1] Homogeneously decomposable modules, Studies on Abelian Groups, Springer-Verlag Dunod (Paris, 1968), 223-238.

[2] A theorem on pure submodules Canadian J. Math. 12 (1960), 438-487.

Krol, M.

[1] The automorphism groups and endomorphism rings of torsion-free abelian groups of rank two, Dissertationes Math. 55 (Warsaw, 1967).

Lady, E.L.

[1] Summands of finite rank torsion free abelian groups, J. Alg. 32 (1974), 51-52.

[2] Nearly isomorphic torsion free abelian groups, J. Alg. 35 (1975), 235-238.

[3] Almost completely decomposable torsion free abelian groups, Proc. A.M.S. 45 (1974), 41-47.

Mal´cev, A.I.

[1] Torsion-free abelian groups of finite rank [Russian] , Mat. Sb. 4 (1938), 45-68.

Monk, G.S.

[1] A characterization of exchange rings, Proc. Amer. Math. Soc. 35 (1972), 344-353.

Murley, C.

[1] The classification of certain classes of torsion free abelian groups, Pac. J. Math. 40 (1972), 647-665.

[2] Direct products and sums of torsion free abelian groups, Proc. Amer. Math. Soc. 3° (1973), 235-241.

Orsatti, A.

[1] A class of rings which are the endomorphism rings of some torsion-free abelian groups, Ann. Scuola Norm. Sup. Pisa 23 (1969), 143-153.

Parr, J.T.

[1] Endomorphism rings of rank two torsion-free abelian groups, J. London Math. Soc. 22 (1971), 611-632.

Pierce, R.S.

[1] Subrings of simple algebras, Michigan Math. J. 7 (1960), 241-243.

Pufter, G.

[1] Antinil groups, Tamkang J. of Math. 6 (1975), 105-108.

Ree, R., and Wisner, R.J.

[1] A note on torsion-free nil groups, Proc. Amer. Math Soc. 7 (1956), 6-8.

Reid, J.D.

[1] A note on torsion-free abelian groups of infinite rank, Proc. Amer. Math. Soc. 13 (1962), 222-225.

[2] On quasi-decompositions of torsion-free abelian groups, Proc. Amer. Math. Soc. 13 (1962), 550-554.

[3] On subgroups of an abelian group maximal disjoint from a given subgroup, Pacific J. Math. 13 (1963), 657-663.

[4] On the ring of quasi-endomorphisms of a torsion-free group, Topics in Abelian Groups, 51-68 (Chicago, Illinois, 1963).

[5] On subcommutative rings, Acta. Math. Acad. Sci. Hungar. 16 (1965), 23-26.

[6] On rings on groups, Pacific J. Math. 52 (1974), 55-64.

Reiner, I.

[1] Maximal Orders, Academic Press, New York, 1975.

Richman, F.

[1] A class of rank 2 torsion free groups, Studies on Abelian Groups, (Paris, 1968), 327-333.

Roggenkamp, K.W.

[1] Lattices over Orders II. Springer-Verlag Lecture Notes No. 142, Springer, Berlin. 1970.

Rotman, J.

[1] Torsion-free and mixed abelian groups, Illinois J. Math. 5 (1961) 131-143.

[2] The Grothendieck group of torsion-free abelian groups of finite rank, Proc. London Math. Soc. 13 (1963), 724-732.

[3] Notes on Homological Algebra Van Nostrand Reinhold Mathematical Studies #26, 1970.

Schultz, P.

[1] The endomorphism ring of the additive group of a ring, J. Aust. Math. Soc. 15 (1973), 60-69.

Swan, R.G.

[1] Projective modules over group rings and maximal orders, Ann. of Math. (2) 76(1962), 55-61.

[2] Algebraic K-theory, Springer-Verlag Lecture Notes No. 76, Springer, Berlin, 1968.

Swan, R.G., and Evans, E.G.

[1] K-theory of finite groups and orders, Springer-Verlag Lecture Notes, No. 149, Springer, Berlin, 1970.

Szekeres, G.

[1] Countable abelian groups without torsion, Duke Math. J. 15 (1948), 293-306.

Vinsonhaler, C.

[1] Torsion free abelian groups quasi-projective over their endomorphism rings II, Pac. J. Math. 74 (1978), 261-265.

[2] Almost quasi-pure injective abelian groups, Rocky Mountain J. Math.

Vinsonhaler, C., and Wickless, W.

[1] Torsion free abelian groups quasi-projective over their endomorphism rings, Pac. J. Math. (2) 70 (1977), 1-9.

Walker, C.P.

[1] Properties of Ext and quasi-splitting of abelian groups, Acta. Math. Acad. Sci. Hungar. 15 (1964), 157-160.

Walker, C.L., and Warfield, R.B. Jr.

[1] Unique decompositions of isomorphic refinement theorems in additive categories, J. Pure and Appl. Alg. 7 (1976), 347-359.

Walker, E.A.

[1] Quotient categories and quasi-isomorphisms of abelian groups, Proc. Colloq. Abelian Groups, (Budapest, 1964), 147-162.

Wang, J.S.P.

[1] On completely decomposable groups, Proc. Amer. Math. Soc. 15 (1964), 184-186.

Warfield, R.B. Jr.

[1] Homomorphisms and duality for torsion-free groups, Math. Z. 107 (1968), 189-200.

[2] A Krull-Schmidt theorem for infinite sums of modules, Proc. Amer. Math. Soc. 22 (1969), 460-465.

[3] An isomorphic refinement theorem for abelian groups, Pacific J. Math. 34 (1970), 237-255.

[4] Extensions of torsion-free abelian groups of finite rank, Arch. Math. 23 (1972), 145.

[5] Cancellation for modules and the stable range of endomorphism rings, to appear.

[6] Exchange rings and decompositions of modules. Math. Ann. 199 (1972), 31-36.

[7] The structure of mixed abelian groups, Abelian Group Theory, Springer-Verlag Lecture Notes #616, 1-38.

Wickless, W.J.

[1] Abelian groups which admit only nilpotent multiplications, Pacific J. Math. 40 (1972), 251-259.

Whitney, H.

[1] Tensor products of abelian groups, Duke Math. J. 4 (1938), 495-520.

Zariski, S.

[1] Commutative Algebra, Van Nostrand, 1958.

Zassenhaus, H.

[1] Orders as endomorphism rings of modules of the same rank, J. London Math. Soc. 42 (1967), 180-182.

INDEX

Vol. 787: Potential Theory, Copenhagen 1979. Proceedings, 1979. Edited by C. Berg, G. Forst and B. Fuglede. VIII, 319 pages. 1980.

Vol. 788: Topology Symposium, Siegen 1979. Proceedings, 1979. Edited by U. Koschorke and W. D. Neumann. VIII, 495 pages. 1980.

Vol. 789: J. E. Humphreys, Arithmetic Groups. VII, 158 pages. 1980.

Vol. 790: W. Dicks, Groups, Trees and Projective Modules. IX, 127 pages. 1980.

Vol. 791: K. W. Bauer and S. Ruscheweyh, Differential Operators for Partial Differential Equations and Function Theoretic Applications. V, 258 pages. 1980.

Vol. 792: Geometry and Differential Geometry. Proceedings, 1979. Edited by R. Artzy and I. Vaisman. VI, 443 pages. 1980.

Vol. 793: J. Renault, A Groupoid Approach to C*-Algebras. III, 160 pages. 1980.

Vol. 794: Measure Theory, Oberwolfach 1979. Proceedings 1979. Edited by D. Kölzow. XV, 573 pages. 1980.

Vol. 795: Séminaire d'Algèbre Paul Dubreil et Marie-Paule Malliavin. Proceedings 1979. Edited by M. P. Malliavin. V, 433 pages. 1980.

Vol. 796: C. Constantinescu, Duality in Measure Theory. IV, 197 pages. 1980.

Vol. 797: S. Mäki, The Determination of Units in Real Cyclic Sextic Fields. III, 198 pages. 1980.

Vol. 798: Analytic Functions, Kozubnik 1979. Proceedings. Edited by J. Lawrynowicz. X, 476 pages. 1980.

Vol. 799: Functional Differential Equations and Bifurcation. Proceedings 1979. Edited by A. F. Izé. XXII, 409 pages. 1980.

Vol. 800: M.-F. Vignéras, Arithmétique des Algèbres de Quaternions. VII, 169 pages. 1980.

Vol. 801: K. Floret, Weakly Compact Sets. VII, 123 pages. 1980.

Vol. 802: J. Bair, R. Fourneau, Etude Géometrique des Espaces Vectoriels II. VII, 283 pages. 1980.

Vol. 803: F.-Y. Maeda, Dirichlet Integrals on Harmonic Spaces. X, 180 pages. 1980.

Vol. 804: M. Matsuda, First Order Algebraic Differential Equations. VII, 111 pages. 1980.

Vol. 805: O. Kowalski, Generalized Symmetric Spaces. XII, 187 pages. 1980.

Vol. 806: Burnside Groups. Proceedings, 1977. Edited by J. L. Mennicke. V, 274 pages. 1980.

Vol. 807: Fonctions de Plusieurs Variables Complexes IV. Proceedings, 1979. Edited by F. Norguet. IX, 198 pages. 1980.

Vol. 808: G. Maury et J. Raynaud, Ordres Maximaux au Sens de K. Asano. VIII, 192 pages. 1980.

Vol. 809: I. Gumowski and Ch. Mira, Recurences and Discrete Dynamic Systems. VI, 272 pages. 1980.

Vol. 810: Geometrical Approaches to Differential Equations. Proceedings 1979. Edited by R. Martini. VII, 339 pages. 1980.

Vol. 811: D. Normann, Recursion on the Countable Functionals. VIII, 191 pages. 1980.

Vol. 812: Y. Namikawa, Toroidal Compactification of Siegel Spaces. VIII, 162 pages. 1980.

Vol. 813: A. Campillo, Algebroid Curves in Positive Characteristic. V, 168 pages. 1980.

Vol. 814: Séminaire de Théorie du Potentiel, Paris, No. 5. Proceedings. Edited by F. Hirsch et G. Mokobodzki. IV, 239 pages. 1980.

Vol. 815: P. J. Slodowy, Simple Singularities and Simple Algebraic Groups. XI, 175 pages. 1980.

Vol. 816: L. Stoica, Local Operators and Markov Processes. VIII, 104 pages. 1980.

Vol. 817: L. Gerritzen, M. van der Put, Schottky Groups and Mumford Curves. VIII, 317 pages. 1980.

Vol. 818: S. Montgomery, Fixed Rings of Finite Automorphism Groups of Associative Rings. VII, 126 pages. 1980.

Vol. 819: Global Theory of Dynamical Systems. Proceedings, 1979. Edited by Z. Nitecki and C. Robinson. IX, 499 pages. 1980.

Vol. 820: W. Abikoff, The Real Analytic Theory of Teichmüller Space. VII, 144 pages. 1980.

Vol. 821: Statistique non Paramétrique Asymptotique. Proceedings, 1979. Edited by J.-P. Raoult. VII, 175 pages. 1980.

Vol. 822: Séminaire Pierre Lelong–Henri Skoda, (Analyse) Années 1978/79. Proceedings. Edited by P. Lelong et H. Skoda. VIII, 356 pages, 1980.

Vol. 823: J. Král, Integral Operators in Potential Theory. III, 171 pages. 1980.

Vol. 824: D. Frank Hsu, Cyclic Neofields and Combinatorial Designs. VI, 230 pages. 1980.

Vol. 825: Ring Theory, Antwerp 1980. Proceedings. Edited by F. van Oystaeyen. VII, 209 pages. 1980.

Vol. 826: Ph. G. Ciarlet et P. Rabier, Les Equations de von Kármán. VI, 181 pages. 1980.

Vol. 827: Ordinary and Partial Differential Equations. Proceedings, 1978. Edited by W. N. Everitt. XVI, 271 pages. 1980.

Vol. 828: Probability Theory on Vector Spaces II. Proceedings, 1979. Edited by A. Weron. XIII, 324 pages. 1980.

Vol. 829: Combinatorial Mathematics VII. Proceedings, 1979. Edited by R. W. Robinson et al.. X, 256 pages. 1980.

Vol. 830: J. A. Green, Polynomial Representations of GL_n. VI, 118 pages. 1980.

Vol. 831: Representation Theory I. Proceedings, 1979. Edited by V. Dlab and P. Gabriel. XIV, 373 pages. 1980.

Vol. 832: Representation Theory II. Proceedings, 1979. Edited by V. Dlab and P. Gabriel. XIV, 673 pages. 1980.

Vol. 833: Th. Jeulin, Semi-Martingales et Grossissement d'une Filtration. IX, 142 Seiten. 1980.

Vol. 834: Model Theory of Algebra and Arithmetic. Proceedings, 1979. Edited by L. Pacholski, J. Wierzejewski, and A. J. Wilkie. VI, 410 pages. 1980.

Vol. 835: H. Zieschang, E. Vogt and H.-D. Coldewey, Surfaces and Planar Discontinuous Groups. X, 334 pages. 1980.

Vol. 836: Differential Geometrical Methods in Mathematical Physics. Proceedings, 1979. Edited by P. L. García, A. Pérez-Rendón, and J. M. Souriau. XII, 538 pages. 1980.

Vol. 837: J. Meixner, F. W. Schäfke and G. Wolf, Mathieu Functions and Spheroidal Functions and their Mathematical Foundations Further Studies. VII, 126 pages. 1980.

Vol. 838: Global Differential Geometry and Global Analysis. Proceedings 1979. Edited by D. Ferus et al. XI, 299 pages. 1981.

Vol. 839: Cabal Seminar 77 – 79. Proceedings. Edited by A. S. Kechris, D. A. Martin and Y. N. Moschovakis. V, 274 pages. 1981.

Vol. 840: D. Henry, Geometric Theory of Semilinear Parabolic Equations. IV, 348 pages. 1981.

Vol. 841: A. Haraux, Nonlinear Evolution Equations- Global Behaviour of Solutions. XII, 313 pages. 1981.

Vol. 842: Séminaire Bourbaki vol. 1979/80. Exposés 543–560. IV, 317 pages. 1981.

Vol. 843: Functional Analysis, Holomorphy, and Approximation Theory. Proceedings. Edited by S. Machado. VI, 636 pages. 1981.

Vol. 844: Groupe de Brauer. Proceedings. Edited by M. Kervaire and M. Ojanguren. VII, 274 pages. 1981.

Vol. 845: A. Tannenbaum, Invariance and System Theory: Algebraic and Geometric Aspects. X, 161 pages. 1981.

Vol. 846: Ordinary and Partial Differential Equations, Proceedings. Edited by W. N. Everitt and B. D. Sleeman. XIV, 384 pages. 1981.

Vol. 847: U. Koschorke, Vector Fields and Other Vector Bundle Morphisms – A Singularity Approach. IV, 304 pages. 1981.

Vol. 848: Algebra, Carbondale 1980. Proceedings. Ed. by R. K. Amayo. VI, 298 pages. 1981.

Vol. 849: P. Major, Multiple Wiener-Itô Integrals. VII, 127 pages. 1981.

Vol. 850: Séminaire de Probabilités XV. 1979/80. Avec table générale des exposés de 1966/67 à 1978/79. Edited by J. Azéma and M. Yor. IV, 704 pages. 1981.

Vol. 851: Stochastic Integrals. Proceedings, 1980. Edited by D. Williams. IX, 540 pages. 1981.

Vol. 852: L. Schwartz, Geometry and Probability in Banach Spaces. X, 101 pages. 1981.

Vol. 853: N. Boboc, G. Bucur, A. Cornea, Order and Convexity in Potential Theory: H-Cones. IV, 286 pages. 1981.

Vol. 854: Algebraic K-Theory. Evanston 1980. Proceedings. Edited by E. M. Friedlander and M. R. Stein. V, 517 pages. 1981.

Vol. 855: Semigroups. Proceedings 1978. Edited by H. Jürgensen, M. Petrich and H. J. Weinert. V, 221 pages. 1981.

Vol. 856: R. Lascar, Propagation des Singularités des Solutions d'Equations Pseudo-Différentielles à Caractéristiques de Multiplicités Variables. VIII, 237 pages. 1981.

Vol. 857: M. Miyanishi. Non-complete Algebraic Surfaces. XVIII, 244 pages. 1981.

Vol. 858: E. A. Coddington, H. S. V. de Snoo: Regular Boundary Value Problems Associated with Pairs of Ordinary Differential Expressions. V, 225 pages. 1981.

Vol. 859: Logic Year 1979–80. Proceedings. Edited by M. Lerman, J. Schmerl and R. Soare. VIII, 326 pages. 1981.

Vol. 860: Probability in Banach Spaces III. Proceedings, 1980. Edited by A. Beck. VI, 329 pages. 1981.

Vol. 861: Analytical Methods in Probability Theory. Proceedings 1980. Edited by D. Dugué, E. Lukacs, V. K. Rohatgi. X, 183 pages. 1981.

Vol. 862: Algebraic Geometry. Proceedings 1980. Edited by A. Libgober and P. Wagreich. V, 281 pages. 1981.

Vol. 863: Processus Aléatoires à Deux Indices. Proceedings, 1980. Edited by H. Korezlioglu, G. Mazziotto and J. Szpirglas. V, 274 pages. 1981.

Vol. 864: Complex Analysis and Spectral Theory. Proceedings, 1979/80. Edited by V. P. Havin and N. K. Nikol'skii, VI, 480 pages. 1981.

Vol. 865: R. W. Bruggeman, Fourier Coefficients of Automorphic Forms. III, 201 pages. 1981.

Vol. 866: J.-M. Bismut, Mécanique Aléatoire. XVI, 563 pages. 1981.

Vol. 867: Séminaire d'Algèbre Paul Dubreil et Marie-Paule Malliavin. Proceedings, 1980. Edited by M.-P. Malliavin. V, 476 pages. 1981.

Vol. 868: Surfaces Algébriques. Proceedings 1976–78. Edited by J. Giraud, L. Illusie et M. Raynaud. V, 314 pages. 1981.

Vol. 869: A. V. Zelevinsky, Representations of Finite Classical Groups. IV, 184 pages. 1981.

Vol. 870: Shape Theory and Geometric Topology. Proceedings, 1981. Edited by S. Mardešić and J. Segal. V, 265 pages. 1981.

Vol. 871: Continuous Lattices. Proceedings, 1979. Edited by B. Banaschewski and R.-E. Hoffmann. X, 413 pages. 1981.

Vol. 872: Set Theory and Model Theory. Proceedings, 1979. Edited by R. B. Jensen and A. Prestel. V, 174 pages. 1981.

Vol. 873: Constructive Mathematics, Proceedings, 1980. Edited by F. Richman. VII, 347 pages. 1981.

Vol. 874: Abelian Group Theory. Proceedings, 1981. Edited by R. Göbel and E. Walker. XXI, 447 pages. 1981.

Vol. 875: H. Zieschang, Finite Groups of Mapping Classes of Surfaces. VIII, 340 pages. 1981.

Vol. 876: J. P. Bickel, N. El Karoui and M. Yor. Ecole d'Eté de Probabilités de Saint-Flour IX – 1979. Edited by P. L. Hennequin. XI, 280 pages. 1981.

Vol. 877: J. Erven, B.-J. Falkowski, Low Order Cohomology and Applications. VI, 126 pages. 1981.

Vol. 878: Numerical Solution of Nonlinear Equations. Proceedings, 1980. Edited by E. L. Allgower, K. Glashoff, and H.-O. Peitgen. XIV, 440 pages. 1981.

Vol. 879: V. V. Sazonov, Normal Approximation – Some Recent Advances. VII, 105 pages. 1981.

Vol. 880: Non Commutative Harmonic Analysis and Lie Groups. Proceedings, 1980. Edited by J. Carmona and M. Vergne. IV, 553 pages. 1981.

Vol. 881: R. Lutz, M. Goze, Nonstandard Analysis. XIV, 261 pages. 1981.

Vol. 882: Integral Representations and Applications. Proceedings, 1980. Edited by K. Roggenkamp. XII, 479 pages. 1981.

Vol. 883: Cylindric Set Algebras. By L. Henkin, J. D. Monk, A. Tarski, H. Andréka, and I. Németi. VII, 323 pages. 1981.

Vol. 884: Combinatorial Mathematics VIII. Proceedings, 1980. Edited by K. L. McAvaney. XIII, 359 pages. 1981.

Vol. 885: Combinatorics and Graph Theory. Edited by S. B. Rao. Proceedings, 1980. VII, 500 pages. 1981.

Vol. 886: Fixed Point Theory. Proceedings, 1980. Edited by E. Fadell and G. Fournier. XII, 511 pages. 1981.

Vol. 887: F. van Oystaeyen, A. Verschoren, Non-commutative Algebraic Geometry, VI, 404 pages. 1981.

Vol. 888: Padé Approximation and its Applications. Proceedings, 1980. Edited by M. G. de Bruin and H. van Rossum. VI, 383 pages. 1981.

Vol. 889: J. Bourgain, New Classes of $\mathcal{L}^p$-Spaces. V, 143 pages. 1981.

Vol. 890: Model Theory and Arithmetic. Proceedings, 1979/80. Edited by C. Berline, K. McAloon, and J.-P. Ressayre. VI, 306 pages. 1981.

Vol. 891: Logic Symposia, Hakone, 1979, 1980. Proceedings, 1979, 1980. Edited by G. H. Müller, G. Takeuti, and T. Tugué. XI, 394 pages. 1981.

Vol. 892: H. Cajar, Billingsley Dimension in Probability Spaces. III, 106 pages. 1981.

Vol. 893: Geometries and Groups. Proceedings. Edited by M. Aigner and D. Jungnickel. X, 250 pages. 1981.

Vol. 894: Geometry Symposium. Utrecht 1980, Proceedings. Edited by E. Looijenga, D. Siersma, and F. Takens. V, 153 pages. 1981.

Vol. 895: J.A. Hillman, Alexander Ideals of Links. V, 178 pages. 1981.

Vol. 896: B. Angéniol, Familles de Cycles Algébriques – Schéma de Chow. VI, 140 pages. 1981.

Vol. 897: W. Buchholz, S. Feferman, W. Pohlers, W. Sieg, Iterated Inductive Definitions and Subsystems of Analysis: Recent Proof-Theoretical Studies. V, 383 pages. 1981.

Vol. 898: Dynamical Systems and Turbulence, Warwick, 1980. Proceedings. Edited by D. Rand and L.-S. Young. VI, 390 pages. 1981.

Vol. 899: Analytic Number Theory. Proceedings, 1980. Edited by M.I. Knopp. X, 478 pages. 1981.